Lecture Notes
in Physics

Edited by H. Araki, Kyoto, J. Ehlers, München, K. Hepp, Zürich
R. Kippenhahn, München, H.A. Weidenmüller, Heidelberg,
J. Wess, Karlsruhe and J. Zittartz, Köln
Managing Editor: W. Beiglböck

314

L. Peliti A. Vulpiani (Eds.)

Measures of Complexity

Proceedings of the Conference, Held in Rome
September 30 – October 2, 1987

Springer-Verlag

Berlin Heidelberg New York London Paris Tokyo

Editors

Luca Peliti
Dipartimento di Scienze Fisiche, Università di Napoli
Mostra d'Oltremare, Pad. 19, I-80125 Napoli, Italy

Angelo Vulpiani
Dipartimento di Fisica, Università dell'Aquila
Piazza dell'Annunziata 1, I-67100 L'Aquila, Italy

ISBN 3-540-50316-1 Springer-Verlag Berlin Heidelberg New York
ISBN 0-387-50316-1 Springer-Verlag New York Berlin Heidelberg

© Springer-Verlag Berlin Heidelberg 1988
Printed in Germany

Printing: Druckhaus Beltz, Hemsbach/Bergstr.
Binding: J. Schäffer GmbH & Co. KG., Grünstadt
2158/3140-543210

PREFACE

It has become fashionable to say that current research in theoretical physics is more and more focusing on complex systems. One has even sometimes spoken of the emergence of a "science of complexity", which should deal with the universal features of complex systems - abstracting from the peculiar aspects of the different systems under investigation.

It is indeed possible to identify a few facts which justify these opinions. In condensed matter theory, for example, the main interest has rapidly shifted from the study of "simple" systems - gases, or solids of regular structure and fixed composition - to that of "complex" systems - such as liquids, amorphous systems, or glasses - whose behavior is dominated by disorder. At the same time the theory of dynamical systems has elected as one of its main subjects of study the characterization of deterministic chaos - i.e., of the phenomenon by which systems described by few degrees of freedom, whose dynamics is determined by a simple-looking evolution equation, exhibit a behavior which is not globally predictable. I should also like to mention the related emergence of a "global" point of view in ecology and the trend affecting immunology, where stress is shifting from the consideration of the immune response - with its associated emphasis on antigen-antibody interactions - to the immune system - with emphasis on the structure of the adiotypic network.

It is not clear, however, if these parallel trends do correspond to increasing interest in a common aspect which could be called "complexity". Dictionaries carry several definitions of the word. "Complex" means: (i) composed of interconnected parts; compound; composite; (ii) characterized by a very complicated or involved arrangement of parts, units, etc.; (iii) so complicated or intricate as to be hard to understand or deal with (*The Random House Dictionary*). In what sense are the subjects just mentioned "complex"? One would be tempted to answer, in meaning (iii). But with this kind of negative definition one would not understand whether there is a true peculiarity in the observed trends - since it is the usual endeavor of science at any time to move on towards the problems which at any given time appear "to be hard to understand or deal with". On the other hand, meanings (i) and (ii) surely do not apply to deterministic chaos - which deals with the complex [meaning (iii)?] behavior of simple systems.

Challenged by the recurring references to the birth of a science of complexity and puzzled by considerations of the kind just explained, we thought to resert to a simple minded (though rather "physicalistic") approach. If complexity could be consistently measured, its place among physical concepts could hardly be disputed. (In fact the physical nature of temperature remained a puzzle for almost two centuries, but its relevance was assured by the existence of reliable experimental methods.) This idea prompted us to call for an International Conference on "Measures of Complexity", which, promoted by the University of Rome "La Sapienza" and with the collaboration of the Cassa di Risparmio di Roma, took place at the Dipartimento di Fisica dell'Università "La Sapienza" from September 30 to October 2, 1987. The present volume contains the Proceedings of the Conference. The question asked to the speaker was to identify a manner of quantitatively characterizing complexity within each of their own disciplines and - if possible - to compare it with the corresponding ones of other disciplines.

We all know that a kind of complexity measure exists in information theory: the Shannon entropy H. Larger or lesser complexity of a message corresponds to smaller or larger values of H. This approach makes sense in the typical environment of information theory, when one considers that a receiver collects a message (one among several possible messages) sent through a noisy channel. But within the approach the most complex message consisting of N symbols belonging to an alphabet of C characters is the completely random sequence. We have the expectation however that the complexity random sequence (in which the character at each point is chosen independently of any other point and with equal probability among the possible ones) is of complexity comparable to the trivial one, composed of identical characters.

One aspect is conspicuously absent from the typical thought experiments of information theory: the meaning of the message. But it is this meaning that we strive to identify when we attempt to decode the messages which Nature sends us. In this sense we can say that a random sequence is one to which we attribute no meaning. If one is able to identify the random and nonrandom components of a phenomenon (or of a message), it is possible to measure complexity by means of the information necessary to reproduce, not the details of phenomenon - but its probability distribution. In this way, e.g., chaotic attractors duly appear more complex than the trivial or the completely random ones. On the other hand the identification of meaning is not obvious when one deals with the characterization of biological structures.

We gratefully acknowledge the Cassa di Risparmio di Roma for the collaboration. We also thank Elena Gagliasso, Roberto Livi and Stefano Ruffo for the friendly help in the organization of the conference. A special thank goes to the secretary of the conference Angela Di Silvestro, Marcella Mastrofini and Rossana De Gregorio for the patient and careful work in preparing the meeting and the proceedings.

Roma, July 1988

Francesco Guerra   Luca Peliti   Angelo Vulpiani

CONTENTS

Preface

A Scientific Discussion

COMPLEXITY AND FORECASTING IN DYNAMICAL SYSTEMS

Peter Grassberger
Physics Department, University of Wuppertal
D - 5600 Wuppertal 1, Gauss-Strasse 20

Abstract:

We discuss ways of defining complexity in physics, and in particular for symbol sequences typically arising in autonomous dynamical systems. We stress that complexity should be distinct from randomness. This leads us to consider the difficulty of making optimal forecasts as one (but not the only) suitable measure. This difficulty is discussed in detail for two different examples: left-right symbol sequences of quadratic maps and 0-1 sequences from 1-dimensional cellular automata iterated just one single time. In spite of the seeming triviality of the latter model, we encounter there an extremely rich structure.

1. WHAT IS COMPLEXITY?

There have always been three major directions along which the frontiers of physics have advanced: towards the very large, towards the very small, and towards the complex.

A central rôle in statistical mechanics, the field of physics dealing traditonally with complex systems, is played by entropy. This rôle has many facets. First of all, entropy is a thermodynamic concept closely related to temperature and not needing any microscopic interpretation. We shall not deal with this aspect here. Secondly, as shown by Boltzmann, it is a measure of disorder or randomness. Finally, according to Szilard and Shannon, it measures an amount of information. What information this is depends on the circumstances.

In dynamical system theory, it is the third aspect of entropy

which is the most important. For a chaotic system, the entropy is the
most direct measure of non-determinacy [1], as it measures the amount
of information which one needs in order to specify a long trajectory:
independently of the particular coding used to describe the trajectory
(provided only it is sufficiently fine), this information increases
~ ht, where t is the time and h is called the Kolmogorov-Sinai (or
"metric") entropy. It is due to this linear increase with time that
chaotic systems are impossible to forecast on the long run: even if we
know the initial state extremely precisely, there will come a time
when this information alone is no longer enough to allow any forecast
[2].

But this information aspect is also valid e.g. in equilibrium
statistical mechanics. There, the entropy is the missing information
needed to specify the microstate if the macrostate is given.

While the entropy is still the central concept in chaotic system
theory, it has been realized more and more during the last years that
it does not tell the whole story. There is a wide spread feeling that
besides entropy or randomness there exists something which many people
would like to call "complexity" [3-6]. It seems that we have to get a
better understanding of what this is if we are to understand better
those systems which are now in the forefront of interest like neural
networks, evolving and learning systems, and chaos.

In mathematics and computer science there exists a quite elaborated theory of complexity (see, e.g., [7]), and it might seem at first
straightforward to apply the concepts developed there to physics. This
is not quite true.

Indeed, the most popular definition of complexity of a string of
symbols, the *algorithmic* or *Kolmogorov-Chaitin complexity* [8], leads
in the cases we are interested in just back to entropy. The Kolmogorov
complexity of a string S of N symbols (with N finally tending to infinity) is defined via the shortest string of bits which can produce S
as an output on a general purpose computer. This definition was made
for strings such as the bits in a computer program computing some well
defined function, or as the digits 3141592653... of π. First of all,
neither the bits of a well written program nor the digits of π can be
"random", as they refer to something very specific. Furthermore, although the digits of π look perfectly random to a statistian (several
millions are known, and they have passed tests for randomness brilli-

antly [9]), the required computer programs are surprisingly short (of length $\propto \log N$ only). Thus "randomness" in a statistical sense and "complexity" via a program length are very different here. The reason for this difference is that it is much easier to write a program which gives the *first* N digits of π than one which gives N consecutive digits starting at some random position. It is essentially the latter which is tested by statistical tests.

This difference does not exist for symbol sequences obtained from dynamical systems, provided the system was time translationally invariant, and the initial configuration was randomly chosen from a stationary distribution. There it is obvious that codings of the sequence can be made efficient only by using correlations which exist between the symbols, and hence both concepts are the same. This applies, by the way, also to two- or three-dimensional patterns generated by spontaneous pattern forming processes. Also there, Kolmogorov complexity is identical to Shannon entropy and its use is, though not wrong, misleading (more precisely, Shannon entropy is the average of the complexity, but in these cases one is only interested in averages anyhow).

If we want therefore a "complexity" which is *not* equivalent to entropy in the cases we are interested in (and, as it seems, most people take it intuitively for granted that such a concept exists), we have to look for something else.

The direction where to look is suggested by computer science. In an admittedly vague sense, we can define:

The complexity of an object (pattern, string, machine, algorithm, ...) is the difficulty of the most important task associated with this object.

For instance, the space complexity of an algorithm is the amount of storage on a general purpose computer which it needs [7] (i.e., the difficulty to implement it), while its time complexity is the time it requires.

The Kolmogorov complexity of a sequence is in particular the difficulty of uniquely specifying the entire sequence, and thus it seems at first sight to agree perfectly with this definition. But specifying a sequence or a pattern is not necessarily the most important task related to it. Much more important might be to "<u>understand</u>" it, i.e. to describe its "<u>meaning</u>". The problem with making the latter into something which a physicist can work with is of course that "meaning" and "understanding" are not well defined concepts, and if we were to pursue this road, we would end up with deep philosophical questions.

A measure of complexity in this spirit is Bennett's *"logical depth"* [10]. The logical depth of a string S is essentially the time needed for a general purpose computer to actually run the shortest program which generates S. For a random string, the time needed to generate S consists essentially of the time needed to read in the specification, and is thus proportional to its length. In contrast to this, a string with great logical depth might have a very short program coding for it, but decoding the program takes very long, much longer than the length of S itself. The prime example of a pattern with great logical depth is presumably life itself. As far as we know, life emerged spontaneously, i.e. with a "program" which was assembled randomly, and which therefore had to be very short. But it has taken some 10^9 years to work with this program until life has assumed its present form.

A more formal example with (presumably) large logical depth is the central vertical column in fig.1. This figure is obtained with one of Wolfram's [4] "elementary" 1-d cellular automata, rule #86. In this cellular automaton, one starts with an infinite horizontal row of "0" and "1", and iterates by adding in each time step another row under

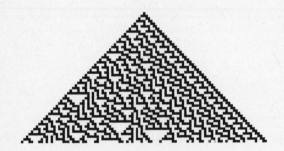

Fig.1: Pattern generated by cellular automaton rule #86, from an initial configuration having one single "1". Time increases downward. The central column seems to be random (after S. Wolfram [11])

the previous one, according to a fixed local rule. In rule #86, one writes "1" under each of the triples 110, 100, 010, and 001, while one writes "0" under every other triple. Figure 1 is got by starting with the row ...0001000..., and by drawing a black square for each "1". Since both the initial configuration and the rule are very easy to describe, the central column has zero Kolmogorov complexity. From very long simulations, it seems however that it has maximal entropy [11], just as the digits of π. Furthermore, it is believed that there exists no other way of getting this column than by direct simulation. Since it takes ~N^2 operations to iterate N time steps, we find that the logical depth is large indeed.

One problem which the logical depth shares with Kolmogorov complexity is that both are not effectively computable. In neither case one can ever be sure to have found the most efficient coding of what may look like a random pattern. Thus, one can only get an upper estimate for the Kolmogorov complexity of some not yet understood pattern found in nature. For the logical depth, this problem is even worse: there, one cannot even be sure wether one's estimate is an upper or lower bound as one could find shorter programs which need either less or more time to execute.

One can avoid this problem be restricting the way in which the pattern is to be encoded. A well known complexity measure obtained in this way is e.g. the *Lempel-Ziv* [12] *complexity* of a string. Here, one uses only the following type of codes: one breaks up the string into "words" which are such that each word consists of a prefix which itself is a word which had appeared previously, plus a suffix which in the simplest version is a single symbol. The coding is done by specifying the suffix and the prefix, the latter again encoded in the same way. As shown by Lempel and Ziv, this gives for any string of length N with a well defined entropy h a coding sequence whose length is asymptotically equal hN, with probability 1. The sequence of binary digits of π, for example, has a Lempel-Ziv complexity of 1 bit/digit. Thus, the Lempel-Ziv measure of complexity is again a measure of randomness rather than the kind of complexity we are looking for. This does not deny that Lempel-Ziv coding and related methods are very useful for many purposes. These include compressing data for electronic storage and transmission [13], or estimating entropies of sequences with long-ranged correlations [14].

A problem with defining complexity in cases which one would intuitively consider as very complex is that such a definition can never be expected to be unique. Indeed, complex situations are characterized by the absense of one single task which can be considered as most important. Take for instance a string of letters such as "The fox is green". One task associated with it is how to tell a computer to print it. This leads to Kolmogorov or Lempel-Ziv complexity, depending on the imposed restrictions. Another task might be to check what language it is written in, if any. After one has found out that it is English, there comes the task of parsing its syntax. A measure for the difficulty of this is the complexity of English grammar. Finally, one has to verify the truth of the statement, whose difficulty is again different (notice that the latter brings us back to the problem of *meaning*). The last tasks (finding the language, parsing the phrase, understandig it, and checking its truth) deal with classifying. This might not be obvious for understanding. But "understanding" the phrase implies that one realizes that it belongs to the set of phrases attributing a colour to an animal, in classifying the kind of animal, etc. The fact that complexity in the intuitive sense is usually related to classifying, or to specifying the ensemble(s) to which the considered object belongs, was stressed in [6].

A last general aspect of complex patterns is that they involve long-range and non-trivial *correlations*. A complex pattern or object cannot be understood simply by describing its parts. Familiar situations in physics are phase transitions, fractal objects like percolation or DLA [15] clusters, or quasicrystals. However, even these correlations are not yet the most complex, as they are too simple due to their scaling invariance. Moreover, in these cases all higher-order correlations are similar to the two-point correlations. The latter is generalized in multifractals [16], but again in a fairly simple and "understandable" way. An example which seems to have much more complex long-range correlations is described in [17].

An interesting class of systems are those with *hierarchical* structure (see, e.g. [3]). Maybe under the impression of spin glasses and self-similar fractals which both show such a structure, several authors have come to consider hierarchies as prototypes of complex structures. We disagree. It is true that hierarchical structures show typical long-range correlations, but as we said these are of a relatively trivial nature. This is maybe most easily seen by considering human societies: both strictly egalitarian and strictly hierarchical societies are ridiculously simple. The complexity of real societies comes from "*tangled*" hierarchies, i.e. hierarchies with internal feed-

back loops. That feedback is an essential ingrediente for complexity
is e.g. evident from nonlinear chaotic systems. It was stressed in
particular by Hofstadter [18].

It seems that the difficulty of the task of determining the most
important task is itself a sign of complexity, in a self-referential
or "Goedelian" way. In order to avoid the problems arising therefrom,
we have to restrict ourselves to relatively simple situations, and
work within a maybe arbitrary but definite scheme. This is what we
shall do in the next section, where we discuss complexity measures
related to forecasting sequences randomly drawn from some ensemble. I
might add that the difficulty of forecasts is closely related to long-
range correlations, in particular when the latter are measured by the
convergence of block entropies. For details, I refer to ref.[6]. In
secs.3 and 4, we shall work out two examples in more details. Conclu-
sions are drawn in sec.5.

2. COMPLEXITIES OF FORECASTING

In the following, we shall avoid the most serious problems by
restricting ourselves to a specific type of sequences and to one specific kind of tasks. More precisely, we consider only sequences made
up of a finite number of different symbols (only 0 and 1, indeed) and
which are outputs of autonomous dynamical systems. That is, they are
strings drawn randomly from ensembles with stationary statistics.

The first example which we shall study in detail is made up of
left-right symbol sequences [19,21] for the quadratic map

$$x_{n+1} = a - x_n^2. \tag{2.1}$$

To each sequence $\{x_n\}$ one can associate a sequence $S = \{s_n\}$ with

$$s_n = \begin{cases} L & \text{if } x<0 \\ R & \text{if } x>0 \\ C & \text{if } x=0 \end{cases} \tag{2.2}$$

For nearly every sequence, the point x=0 will never be reached exactly, and thus nearly every sequence can be encoded by a binary string.

The other example consists of strings produced by 1-d cellular automata (such as fig.1), but with random initial configurations. In particular, we shall consider the horizontal (spatial) strings after a single iteration [20]. This might seem a trivial problem, but as we shall see it is not at all.

The task which we consider as most important is that of forecasting the sequence. Notice that forcasting a sequence is not the same as specifying it. Assume we know exactly the statistics and the "grammar" of the sequence. We first try to predict s_1 without knowing anything about the specific sequence we are dealing with. After we are told the true s_1, we predict s_2 *using that we know* s_1, then we are told s_2 and have to predict s_3, etc. Since the sequences we are interested in have positive entropy, we will never be able to make perfect forecasts. Instead, there will always be an uncertainty of at least h bits per forecast. An optimal forecasting strategy, where this limit is reached asymptotically, might be very easy or very difficult. It is this difficulty which we call the forecasting complexity, and which we consider as the most natural complexity measure of this sequence.

The crucial point to notice is that this difficulty is *not* related to the value of h. For a completely random sequence (such as the R-L sequence of eq.(1) with a=2) entropy is maximal but the *optimal* forecast is very easy: it is just a pure guess. On the other hand, for a quasiperiodic sequence such as a symbol sequence of a circle map at a noble critical winding number, or for the R-L sequence of eq.(1) at the Feigenbaum point a = 1.401155... , the entropy is zero but an optimal forecast requires infinite resources, as we shall see below.

Indeed, the above "definition" does not yet specify the forecasting complexity. We still have to say how we measure the difficulty, what we really want to forecast, and what tools we are allowed to use.

a) Regular language complexity (RLC)

Here, we forget about all probabilistic aspects, and "forecasting" means that we only predict which symbols *can* and which *cannot* appear next. Thus we are only interested in "grammatical" questions when viewing the string as belonging to some formal language. Measures for the difficulty of this task are either the maximal space needed during such a scan, or the maximal time needed for one symbol. Within the Chomsky hierarchy [7], the only class of grammars where both are finite are the "regular languages". For the other Chomsky classes, the maximal space is unbounded. The maximal time diverges then also since already the time needed to address the memory becomes infinite.

Regular languages are such that the grammatical correctness can be verified by means of a finite directed graph. In this graph, each link is labeled by a symbol, and each symbol appears at most once on all links leaving any single node. Furthermore, the graph has a unique start node. Any grammatically correct string is then represented uniquely by a walk on the graph, while any wrong string is not. Scanning the string consists in following the walk on the graph, and forecasting is done by reading the possible labels of the next link. The main difficulty here is that one has to remember the number of the present node. If the smallest graph doing the job for a particular language has n nodes, then this needs $\log_2 n$ bits. The required time is essentially that for reading this number from a table, and is thus also $\sim \log n$. For this reason, $\log n$ was called the *regular language*

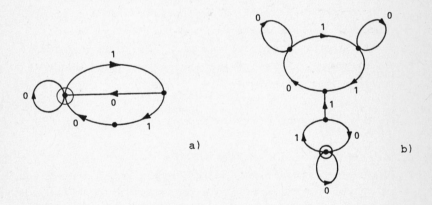

Fig.2: Deterministic graphs for the regular languages generated by rules #76 (a) and 18 (b) (from ref.[5]). The heavy nodes are the start nodes.

complexity by Wolfram [5] (in ref.[6], the RLC was unfortunately called "algorithmic complexity". We use here the name given by Wolfram, in order to avoid confusion with Kolmogorov complexity).

For chaotic R-L symbol sequences of the quadratic map, the RLC is infinite except when the kneading sequence ends periodically. Thus these sequences do not form regular languages in general. It is not known to which of the other Chomsky classes they do belong.

For spatial strings produced by 1-d cellular automata after a finite number of iterations, it was shown by Wolfram [5] that the RLC is finite. The graphs for the 256 "elementary" rules (3-site neighbourhood) after 1 iteration have between 1 and 15 nodes. The graphs for some of the rules are given in fig.2. For the more complex rules their sizes increas very fast with the number of iterations.

b) Set complexity (SC)

In most cases, one has a measure on the set of grammatically correct sequences. Assume that we still are interested only in predicting the wrong and the possibly correct symbols, but we want to use a strategy which takes probabilities into account in order to be most efficient. As we said above, the number of nodes of the graph of a regular language was a good complexity measure since its logarithm measures both the stored information about the sequence and the time needed for a prediction in the worst case. If we are given probabilities, then it seems natural to replace the worst case values by averages. Assume that during a scan of a typical string, node i is visited during a fraction p_i of the time. Then in an optimal strategy one would have to store an information of

$$SC = - \sum_i p_i \log_2 p_i \qquad (2.3)$$

bits about the past history of the sequence, and the mean time needed to get the next symbols from the table is of the same order of magnitude. Under the given premises, it is then natural to consider SC as the most relevant complexity measure.

One might object that in addition to the information about the scanned sequence one has to store the information about the topology of the graph. This needs an information proportional to its size, much larger than SC, and should thus be considered as the most important. But this would not take into account that (i) most of this information can be stored on slow and inexpensive carriers as it is rarely used, and (ii) this information can be used by many users who share the same computer or computer network. It is not the amount of stored information which is most costly, but the amount of transferred information.

For cellular automaton – generated sequences, it was found in [5] that the SC is for some rules much smaller than the RLC. In these cases, different parts of the graphs are visited with very unequal probabilities. In particular, one can even have finite SC with an infinite accepting graph. The latter seems to happen also for R-L sequences of eq.(2.1), in general.

c) <u>Forecasting complexity proper (FC)</u>

Finally, let us consider the case where one not only wants to check grammatical correctness, but where one also wants the probabilities for the next symbols. Again we take the average (Shannon) information about the past history as the most appropriate complexity measure. In ref.[5], it was called "true measure complexity", but we propose here to call it instead the forecasting complexity (FC). Clearly, one has always $FC \geq SC$.

Denote by $p(i|S_n)$ the forecasted probability to find symbol i in the (n+1)-st position of the sequence, after having already observed the sequence $S_n = (s_1, \ldots s_n)$. In cases where the set of all $p(i|S_n)$ is infinite, the FC is infinite too. In such cases, one has first to introduce some "coarsegraining" approximation, and study how the FC diverges when the coarsegraining is removed.

We shall in the following consider only the case where the $p(i|S_n)$ are countable. In this case, it is often again useful to use a (possibly infinite) graph. In this graph, each link is not only labeled by a symbol, but it carries also the forecasted probability $p(i|Q)$. Here, Q is the node from which the link leaves, and the needed information about the sequence S_n is "encoded" in the number of this node. In this case, one has

$$ h = - \sum_Q p(Q) \sum_i \log_2 p(i|Q) \tag{2.4} $$

and

$$ FC = - \sum_Q p(Q) \log_2 p(Q) \quad . \tag{2.5} $$

In the following section, we shall dicuss L-R sequences of eq.(2.1) in more detail, and in sec.4 we treat 0-1 sequences of cellular automata after 1 iteration.

3. SC FOR THE QUADRATIC MAP

For quadratic maps (indeed, for all continuous maps of an interval onto itself with a single maximum), we start from well known facts [19,21]. First, the "kneading sequence" K is the R-L itinerary of the critical point (starting, for a>0, with "R"). For any R-L sequence $S = (s_1, s_2, \ldots)$, we denote by $\sigma(S)$ the shifted sequence $(s_2, s_3, \ldots)$, and by $\tau(S)$ the sequence $(t_1, t_2, \ldots)$ where $t_1 \varepsilon \{0,1\}$ is the number (modulo 2) of "R"'s in S up to position i. For a = 1.8, e.g., we have $K = RLLRLRRL\ldots$ and $\tau(K) = 11100100\ldots$. Finally, we order the $\tau(S)$ as if they were binary representation of real numbers in [0,1].

A symbol sequence for any one-humped map is allowed iff [21]

$$1 - \tau(K) \leq \sigma^m(\tau(S)) \leq \tau(K) \qquad \text{for all } m \geq 0. \qquad (3.1)$$

This means that the set of allowed sequences is characterized by a list (in general infinite) of forbidden "words". In the above example, the first forbidden words in $\tau(S)$ are 1111 and 0000, and they correspond to the forbidden word LLL in S. The next forbidden words in S are in this example LLRR, LLRLRL, and LLRLRRR.

There is always at most one forbidden word of given length N, and this word is obtained by dropping the first letter from K and exchanging its (N+1)-st letter. For any N, it is easy to give a graph whith $\leq$ N nodes which accepts exactly those sequences which have no forbidden word of length $\leq$ N. Notice that this is not entirely trivial. For a regular language with forbidden words of length up to N one knows in general only that the smallest accepting graphs has to have between N and 2^{N-1} nodes.

These graphs can be constructed by a very simple algorithm [22]. The first few graphs for the above example are given in fig.3. For any

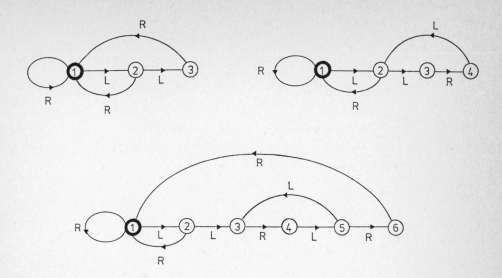

Fig.3: Graphs accepting all L-R sequences for x' = 1.8-x² which
contain no forbidden words of length ≤3, ≤4, and ≤6. These
words are LLL, LLRR, and LLRLRL (from ref. [21]).

N, the graph associates a topological Markov chain to the dynamical
system (2.1). Usually, such an association is made by a partitioning
of state space. In the case of one-humped maps, such partitionings
seem rather non-trivial [23]. It is mainly taking into account the
existence of a start node (something which is usually not done in sym-
bolic dynamics!) which simplifies things.

For N ->∞, and after eventual minimizations, one gets finite
graphs (i.e., regular languages) in periodic windows and at band-
merging points [22]. In all other cases, the graphs are infinite. In
all "typical" chaotic cases the SC seems to be finite, indeed the sum
(2.3) in seems to converge exponentially there [22]. It is not known
to what Chomsky class the languages belong there. At Feigenbaum points
and below tangent bifurcation points, the sum in (2.3) diverges, and
the SC is infinite. This reflects the long-range correlations there
which make renormalization group treatments appropriate.

The graph of SC versus the control parameter a is given in fig.4,
together with the familiar bifurcation diagram. We clearly see the di-
vergence at the beginnings (tangent bifurcation) and endings (Feigenbaum points) of periodic intervals.

Thus we have found that the tasks of verifying the "grammatical" correctness of symbol sequences, and forecasting forbidden symbols, require in general finite efforts when measured via the SC. While the maximal information needed to be stored for this task is infinite, the average amount of information needed is finite. The task of actually forecasting the probabilities with which the allowed symbols appear would, in contrast, require infinite efforts in average, for typical chaos. Details have not yet been worked out there. Instead, for this task we shall switch to the simpler case of cellular automata.

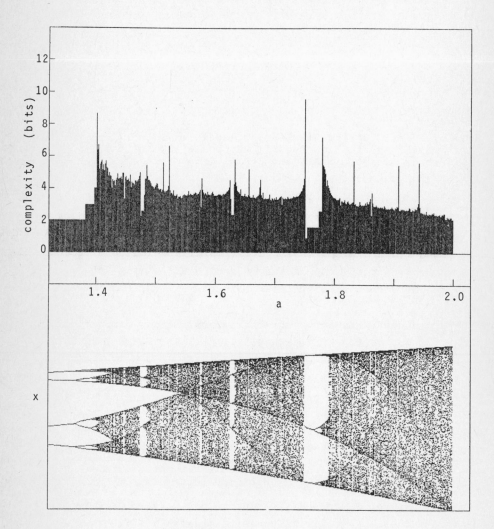

Fig.4: Set complexity of the R-L symbol sequences of eq.(2.1) versus the parameter a, together with the bifurcation diagram.

4. ONE-TIME-STEP CELLULAR AUTOMATA

We assume that in our 1-dimensional cellular automata the input is a random string $T = \ldots t_{n-1}, t_n, t_{n+1}, \ldots$ with $t_n \in \{0,1\}$. The output string S has then

$$s_n = F(t_{n-1}, t_n, t_{n+1}) \tag{4.1}$$

where $F(t,t',t'')$ is any of the 256 boolean functions of 3 arguments numbered in Wolfram's [4] notation. For any F (any "rule"), the language of all S is regular, with the graphs for some rules given in fig.2. As was said, the number of nodes in such a graph gives the information needed to forecast which symbol is forbidden in the next step.

The optimal strategy of actually forecasting S is given in [20]. Call $P_n(t,t')$ the probability that in the input string $t_{n-1} = t$ and $t_n = t'$, *conditioned* on the output string $s_1 \ldots s_{n-1}$. For n=1, there is not yet any observed output string, and thus $P_1(t,t') = \frac{1}{4}$. We call $P_n(t,t')$ the "conjecture" about the input string. We are not directly asked to make this conjecture, but we need it for the required forecast of $p_n(s)$ which is the conditional probability that $s_n = s$:

$$p_n(s) = \frac{1}{2} \sum_{t,t',t''} P_n(t,t') \, \delta[s-F(t,t',t'')] . \tag{4.2}$$

Here, $\delta[i,k]$ is the Kronecker delta. After we have learned the actual s_n, we can update our conjecture about the input string with

$$P_{n+1}(t',t'') = [p_n(s_n)]^{-1} \sum_t P_n(t,t') \, \delta[s_n - F(t,t',t'')] , \tag{4.3}$$

use this to forecast s_{n+1}, go on observing s_{n+1}, etc. Introducing a compact vector notation with

$$P_n = (P_n(0,0), P_n(0,1), P_n(1,0), P_n(1,1)) \tag{4.4}$$

a 4-dimensional vector, we can write eq.(4.3) in matrix form as

$$P_{n+1} = [p_n(s_n)]^{-1} M(s_n) P_n . \tag{4.5}$$

Equations (4.2) and (4.3) together can be considered as a dynamical system with state space consisting of the "conjectures", and with a random input consisting of the string S. It is similar to the "iterated function systems" of ref.[24] except for two points: the division by $p_n(s_n)$ keeps the normalization correct, and the "input" string S is not completely random. In order to make an optimal forecast, we have to simulate this system in parallel with the original system, and this is what makes forcasting difficult if the number of different conjecture vectors is large.

This number depends on the CA rule. All conjecture vectors are again naturally arranged as nodes of a directed graph. The start node is always the vector $P_1 = (1,1,1,1)/4$. The graph for rule #76 is given in fig.5. In this figure, we have a transient infinite part. After the first observation of two "1" in succession, only the finite lower part is relevant. Notice that even the latter part is more complicated than the graph in fig.2a.

The forecasting graphs for all 256 rules have been studied in [20]. While there are very few rules with a finite graph, there are many

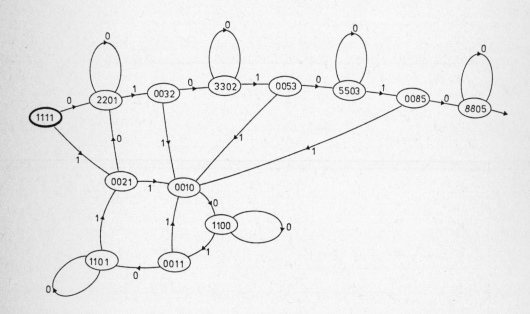

Fig.5: Minimal deterministic graph needed to forecast the CA rule #76 with random input string. Each node is labelled by a "conjecture" $P \in Q^4$ (given here by 4 integers, after having multiplied the $P_n(t,t')$ by the smallest common multiple of their denominators). The actual forecasts are obtained with eq.(4.2).

with an infinite transient part and a finite rest such as the above [20]. But there exist also a large number of rules with an infinite non-transient part. In some of them, this is a simple linear chain as in fig.5, but in the more complicated cases this is no longer true. As an example, we show in fig.6 part of the graph for rule #22. We have not been able to detect any structure in this graph. The number

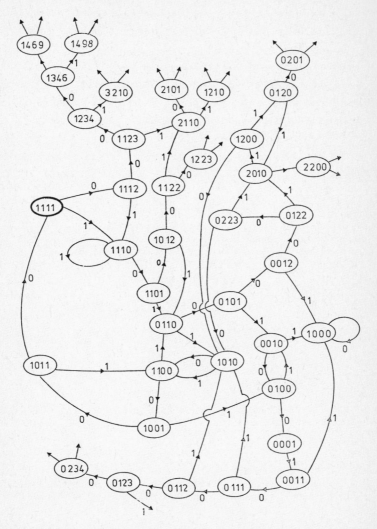

Fig.6: Part of the (presumably infinite) graph for forecasting rule #22 (from ref.[19]).

of nodes with distance ≤ n from the start grows as $e^{(0.386 \pm .004)n}$, as found from exact enumerations with n ≤ 24. The number of different forecasts $p_n(0)$ grows in the same way. Both the set of different conjectures and the set of forecasts seem to be fractals (see fig.7), the latter seems to be dense in the interval [0,½]. However, one can prove that the FC is finite, and the sum in eq.(2.3) converges actually exponentially. The reason is that there exist a "resetting string" [20]. This is a finite string which is produced with non-zero rate in the output. Whenever it does appear, it leads one back to the start node regardless where one was before on the graph. Thus, all node probabilities have to decrease exponentially with their distance from the origin. With minor modifications, the same holds for all other rules.

This means that for all 1-dimensional CA of this type one can make optimal forecasts with finite average effort, though the maximal amount of information to be stored is again infinite.

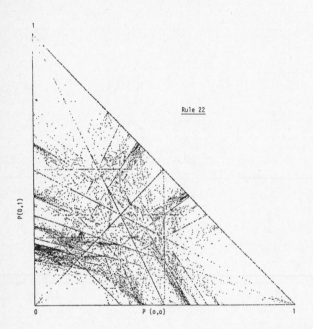

Fig.7: Set of conjecture vectors for rule #22. Shown is the projec-
tion onto the (P(0,0),P(0,1))-plane. The original set seems
to have fractal (box counting) dimension ~2.2 (from ref.[19]).

In ref. [20], we studied also the effect of approximations leading to non-optimal forecasts. For instance, we can tolerate an error $\pm\varepsilon$ in the forecasts $p_n(s)$. The decrease of stored information should be related to the information dimension of the set of forecast vectors. As it turns out, the information dimension is zero (in contrast to the box counting dimension!). We shall not go into detail her, but refer the interested reader to ref.[20]. We just mention that it seems to be fairly easy in these examples to make nearly optimal forecasts, in that the errors converge in general exponentialy with the effort. This is presumably not so in really very complex cases as the one studied in ref.[17].

5. CONCLUSIONS

We have first seen that one should not expect a universal notion of complexity applicable to all situations. This is in contrast to the case of randomness which is measured by entropy. Our premise here was that there exists an intuitive notion of complexity *different* from that of randomness (it is complexity which we mean when we say e.g. that a neural network is a prototype of a complex object).

Generally speaking, we have to define the complexity of some "object" as the difficulty of some task associated with the object. But complex situations are characterized by not having a single most relevant task associated with them, and it is not clear what tools finally might be needed. So we have to chose arbitrarily what we consider the most relevant, and we have to restrict arbitrarily the tools to be used.

While the notion of Kolmogorov complexity of a symbol string is derived from the task of specifying the string in a way suitable to reproduce it on a general purpose computer, we claimed that for the kind of strings generated by dynamical systems at least two other tasks are more important: that of actually constructing the string from its "blueprint" (leading to Bennetts "logical depth"), and that of forecasting the string. Both are closely related in that they can involve an effort much larger that $O(N)$ for a string of length N. Also, this effort is in both cases due to the necessity to understand the rules for the string (or, equivalently, the ensemble from which it is drawn).

While the forecasting complexity is just the usual complexity of the "grammar" if probabilities are neglected, we argued that in our cases it is very important to take probabilities into account. Depending on the specific task, we ended up with two main complexity measures which we studied then in more detail for two specific (classes of) examples.

Both examples are essentially toy models. In particular the latter
might have seemed a priori to be too trivial to be interesting. We
found that it was not at all. Instead we found very rich structures
which are only partially understood yet. I hope that studying such
models will be ultimately useful in understanding "real" problems,
though we are still far from that.

Our results can also be viewed from two different angles. They show
again that life can be made very difficult if there are hidden
observables such as the input string T in sec.4. Even though T was a
trivial Markov chain, and the output string S a deterministic function
thereof, it required considerable effort to forecast S when T is not
observed. The other aspect is that it suggests close relations between
strings without a probability on them (formal languages) and strings
carrying a probability. Markov chains form there only the very
simplest possibility. They are analogous to finite complement languages, a simple subclass of the already simple class of regular
languages [7]. Does for probabilistic strings exist something analogous to the Chomski classification? It seems that not much is known.

The material presented in sec.4 was obtained in collaboration with
Domenico Zambella. I want to take here the opportunity to thank him
for this most pleasant collaboration.

REFERENCES

1. J.-P. Eckmann and D. Ruelle, Rev. Mod. Phys. **57**, 617 (1985)

2. R. Shaw, Z. Naturforsch. **36a**, 80 (1981)

3. T. Hogg and B.A. Huberman, Physica **22D**, 376 (1986);
   C.P. Bachas and B.A. Huberman, Phys. Rev. Lett. **57**, 1965 (1986);
   H.A. Cecatto and B.A. Huberman, Xerox preprint (1987)

4. S. Wolfram, Rev. Mod. Phys. **55**, 601 (1983)

5. S. Wolfram, Commun. Math. Phys. **96**, 15 (1984)

6. P. Grassberger, Int. J. Theoret. Phys. **25**, 907 (1986)

7. J.E. Hopcroft and J.D. Ullman, *Introduction to Automata Theory,
   Languages, and Computation* (Addison-Wesley, New York 1979)

8. A.N. Kolmogorov, *Three Approaches to the Quantitative Definition of Information*, Probl. Inform. Transmiss. 1, 1 (1965);
   G. Chaitin, J. Assoc. Comp. Mach. 13, 547 (1966)

9. S. Wagon, Mathem. Intell. 7, 65 (1985)

10. C.H. Bennett, in *Emerging Syntheses in Science*, D. Pines editor, 1985

11. S. Wolfram, *Random Sequence Generation by Cellular Automata*, to appear in Adv. Appl. Math.

12. A. Lempel and J. Ziv, IEEE Trans. Inform. Theory 22, 75 (1976);
    J. Ziv and A. Lempel, IEEE Trans. Inform. Theory 23, 337 (1977);
    24, 530 (1978)

13. T.A. Welch, Computer 17, 8 (1984)

14. P. Grassberger, preprint (1987), subm. to IEEE Trans. Inform. Theory

15. T.A. Witten and L.M. Sander, Phys. Rev. Lett. 47, 1400 (1981)

16. G. Parisi, appendix in U. Frisch, *Fully Developed Turbulence and Intermittency,* in Proc. of Int. School on *"Turbulence and Predictability in Geophysical Fluid Dynamics and Climate Dynamics"*, M. Ghil editor (North Holland, 1984);
    R. Benzi et al., J. Phys. A17, 3521 (1984)

17. P. Grassberger, J. Stat. Phys. 45, 27 (1986)

18. D.R. Hofstadter, *Gödel, Escher, Bach: an Eternal Golden Braid* (Vintage Books, New York 1980)

19. P. Collet and J.-P. Eckmann, *Iterated Maps on the Interval as Dynamical Systems* (Birkhauser, Basel 1980)

20. D. Zambella and P. Grassberger, preprint (march 1988)

21. J. Dias de Deus, R. Dilao, and A. Noronha de Costa, Lisboa preprint (1984)

22. P. Grassberger, preprint WU-B 87-5 (1987)

23. F. Hofbauer, Israel J. Math. 34, 213 (1979); 38, 107 (1981);
    Erg. Th. & Dynam. Syst. 5, 237 (1985);
    P. Collet, preprint (1986)

24. J.E. Hutchinson, Indiana Univ. Math. J. 30, 713 (1981);
    M.F. Barnsley and S. Demko, Proc. Royal Soc. London A399, 243 (1984)

On Complexity.

Giorgio Parisi

II Universita' di Roma "Tor Vergata",
Dipartimento di Fisica
and
INFN, sezione di Roma.

Abstract

How to define complexity? How to classify the configurations of a complex system? Which are the main features of such a classification? These and similar problems are briefly discussed in this talk.

The definition of complexity is not an easy task, practically each of the speakers has used a different definition, which may range from the classical algorithmic complexity, to more recent and sophisticated definitions. Some times a complex system is defined in more general terms: a complex system is complicated system, composed of many parts, whose properties cannot been understood. It is clear such a given definition cannot capture all the complex meaning we associate to the word complexity, however I will try to present a (may be different) definition, expanding an older proposal of mine[1].

The basic idea is that more the system is complex, more you can say something about it. Of course I am excluding the factual description of the system, which may be very long; I refer to global characteristics of the system. A few examples help in making this point clear. If I have a sequence of randomly tossed coins, 50% probability head, I have already described the system at my best, the only improvement would be the description of the sequence. If on the contrary the sequence of bits represents a book, there are many many things that I can say on the style, the choice of the words, the plot and so on. If the book is really deep, complex, there are many many things you can say about it. Sometimes the complexity is related to the existence of different levels of descriptions: you can describe an Escherichia Coli at the atomic level, the biochemical level and the functional level.

If we try to move toward a mathematical definition, we must realize that the concept of complexity, like entropy, is of probabilistic nature and it can be more precisely defined if we try to define the complexity of ensembles of objects of the same category. Of course, if you have only one object which changes with the time, you can study the complexity of the time dependence (or the behavior) of this object.

The simplest situation for which we can give a consistent definition of complexity, arises when we have an ensemble of sets and we try to classify them. All of us should have some experience in classification; indeed one of the main activity of the mammal mind consists in finding relations among the extremely large amount of sensory information and in classifying them: for example different images of the same object are usually correctly classified as different images of the same object.

Although there is a freedom in deciding how different objects should be classified, we may hope that the qualitative features of the best classification we can do depend on the properties of the external world. We are led to consider the following very general problem: we have a sequence of configurations C's which are generated according to given algorithm A; the algorithm A is not a specific algorithm, but it is a generic algorithm belonging to a given class of algorithms. We would like to classify these configurations in the best possible way.

In order to be more precise on what I mean by the word classification, let us consider some examples: if the configurations are a sequence of completely random numbers, no classification is possible and all the configurations belong to the same class; if we consider the equilibrium configurations of a ferromagnetic system at low temperatures and at zero magnetic field, we can divide them into two sets according if the large majority of spins point in the up or down direction; if our configurations are the living objects on the earth, the best classification is likely the one done in biology and zoology; in history of arts we could classify the different painters by the relative influence of one painter on the other.

We should notice that in the first three examples the configurations are classified as the leaves of a tree (taxonomy), the tree being trivial in the fist two cases; in the last example the situation is more complex and a simple genealogical tree cannot be established (a given painter may be under the influence of many painters). In our mind classification is equivalent to establish some relations of kinship (or distance) among different configurations and the final output is in general is not a tree.

Our aim is not only to establish a classification of the configurations arising from a given algorithm, we want to find which characteristics of the classifications are universal (i.e. they are the same for all the generic algorithms of a given class) and which characteristics depend on the given algorithm; moreover we would like to know the probability distribution of those characteristics which depend on the algorithm. In studying cellular automata, non equilibrium statistical mechanics (and may be biology), we find that quite complex structures arise: we badly need a theory of complexity in order to make progresses in these fields and the program we have outlined may be the starting point.

If we remain at this level of generality, the problem would be hardly soluble. It is convenient to study this problem in a simple framework, i. e. equilibrium statistical mechanics. The results I will outline have been obtained in the framework of equilibrium statistical mechanics and may be considered the first steps toward the construction of the general theory of classifications[2]. These results have been obtained originally in the study of the mean field approximation for spin glasses[3], but it is evident that their interest goes much beyond the specific field of spin glasses.

At it is usual, more precise and cleaner results may be obtained when the dimensions of the configuration space goes to infinity; as in thermodynamics we are led to consider a system in a box of volume V, with V very large (sharp results are obtained in the limit where V goes to infinity).

Let me now review a well known theorem[4] in the framework of equilibrium statistical mechanics on the classification on equilibrium configurations, using a slightly unusual language (the theorem has be proved for translational invariant Hamiltonians, while the case on non translational invariant Hamiltonians is the most interesting one).

The first thing we do is to introduce a topology in the space of configurations by defining an appropriate distance. For simplicity we suppose that our configurations are scalar fields $\varphi(x)$, which are smooth functions of the coordinate x (we could also have considered field defined on a lattice). A natural definition of the distance between two configurations φ_1 and φ_2 is the following:

(1) $d = 1/V \int dx \ |\varphi_1(x) - \varphi_2(x)|^2$.

where V is the volume of the region where the fields φ are defined (at the end V will go to infinity). Using this definition of distance, two configurations, which differs only in a finite region. are identified (they are at distance zero when V goes to infinity).

Different definitions of the distance may be used: for example we could define:

(2) $d = 1/V \int dx \ |\Delta\varphi_1(x) - \Delta\varphi_2(x)|^2$.

General speaking we could use as a definition of a distance:

(3) $d = 1/V \int \ |O[\varphi_1(x)] - O[\varphi_2(x)]|^2$

where O [φ] is a local operator, i. e. it depends only on φ and its derivatives.

We have two alternatives: or we use a definition of distance (e.g. (1)) or we define a distance vector and the true distance will be the norm of the vector; for example the use of a vectorial distance can be useful in biology, if we need to compare the distance between various morphological characters of various species.

If we consider the case of equilibrium statistical mechanics, the probability distribution of the fields φ is

(4) $P[\varphi]= \exp (-\beta \int dx \; H[\varphi]) / Z,$

where the Hamiltonian H is a local (or quasi local) operator (i. e. $\delta^2 H/\delta\varphi(x)\delta\varphi(y)$ goes to zero very fast when $|x-y|\to\infty$) and Z is a number (the partition function) such that the total probability P is normalized to 1 (Boltzman Gibbs probability distribution).

The algorithm we consider here sorts the configurations φ according to the probability distribution eq. (4). A well known theorem[4] (under the appropriate qualifications) states that we can divide the configuration space into disjoint sets labeled by α such that the distance (using an arbitrary definition of the distance in agreement with eqs.(1-3)) between two elements of different sets (e.g. α and γ) does not depend on the elements (we neglect a part of the configuration space whose weight goes to zero when V goes to infinity); more precisely if $\varphi_1 \in S_\alpha$ and $\varphi_2 \in S_\gamma$ ($\varphi_1 \neq \varphi_2$) the distance of φ_1 and φ_2 is a function of α and γ only(which we call $d(\alpha,\gamma)$.

The sets labeled by α can be called phases of the system (or species if we use a biological terminology); as an example we can consider water at zero centigrade and classify its equilibrium configurations as solid or liquid.

In the same way we can decompose the probability distribution eq.(4) as follows:

(5) $P[\varphi]= \sum_\alpha w_\alpha P_\alpha[\varphi],$

where the P_α's are normalized probability distributions concentrated on the set S_α's and the w's satisfy the obvious relation:

(6) $\sum_\alpha w_\alpha=1.$

If we define by $< >$ and by $< >_\alpha$ the expectation value with respect to the probabilities $P[\varphi]$ and $P_\alpha[\varphi]$ respectively, eq. (5) can be written as

(7) $< > = \sum_\alpha w_\alpha < >_\alpha.$

The theorem we have stated says that any equilibrium state ($< >$) can be decomposed as the linear convex combination of pure clustering states ($< >_\alpha$), a clustering state being defined by the property that the connected correlations functions go to zero at large distances. Indeed an easy computation show that the distance is independent on the configurations only if the connected correlations functions go to zero at large distances.

In this case the only thing that can be globally said about the classification are the w's and the distances d, a possible measure (C[w]) of the complexity of the classification could be given by:

$$(8) \qquad C[w] = -\sum_\alpha w_\alpha \ \log(w_\alpha),$$

i. e. the entropy of the set of phases. This definition of complexity is very simple minded and does not capture all the possible variations which may be present in the sets of w's and d's.

We have thus arrived to the conclusion that in equilibrium statistical mechanics only the phases of the system must be classified. This task is normally simple for many translational invariant Hamiltonians, but it may be rather complex for random system like spin glasses.

In spin glass the only solid results have been obtained in the mean field approximation (which should be valid when the dimensions D of the space are very large or in presence of long range forces) an highly non trivial structure arises[4]. The different pure states may be grouped into clusters: all elements of a cluster have the same distance among them and elements of two different clusters stay at a distance which does not depend on the elements we consider; clusters may be grouped into superclusters with similar properties, supercluster may grouped into supersuperclusters and so on. In other words the phase of the system are the leaves of tree which is hierarchically organized with infinite many levels; the statistical properties of the w's can be studied in great detail.

In the case where the configurations may be organized on a tree in such a way that the distance among two configurations depends on the position on the tree, the space of configuration is ultrametric and the distance satisfies the ultrametricity inequality:

$$(9) \qquad d(\alpha,\gamma) < \max \ (d(\alpha,\delta) \ , \ d(\delta,\gamma)) \qquad \forall \ \delta.$$

The ultrametricity property corresponds to the simplest possible non trivial organization of the states and it is quite possible that it will be present also in optimization problems like the traveling salesman or the matching problem[5]. An open problem is to understand if and how the ultrametricity property breaks down when we decrease the dimensions D of the space and if more complex distributions may be generated; in general we do not know how to cope with these more complex distributions, e.g. how we should parametrize them.

The results obtained in the theory of spin glasses suggest that the study of complex system should be divided into two steps.

We first start from the microscopic (low level) description of the system and we compute the probability distribution $P\{w,d\}$ of the weights and of the distances of the various pure phases of the system. The complexity could be defined (in a more appropriate way than in eq. (8)) as

$$(10) \qquad C = -\int d\{w,d\} \ P\{w,d\} \ \log[\ P\{w,d\} \ ].$$

The second step (high level statistic mechanics) should consist in computing the various statistical properties of the distances using the function P{w,d} as a starting point. We are just at the beginning of this ambitious program and it is clear that it will be successfully only if the form of the function P{w,d} will be an universal quantity (like the critical exponents for second order phase transitions) in the sense that it will not depend on the fine microscopic details.

References

1) G. Parisi Physica Scripta, 35, 123 (1987).
2) M. Mezard, G. Parisi, N. Sourlas, G. Toulouse and M. Virasoro, Phys. Rev. Lett. 52, 1156 (1984); J. Physique 45, 843 (1984), M. Mezard, G. Parisi, M. Virasoro, Europhys. Lett, 1, 56, (1986).
3) A theoretical review of spin glasses and related subjects can be found in G. Parisi, in "Field Theory and Statistical Mechanics", ed. by J. B. Zuber and R. Stora, North Holland (1984). and in M. Mezard, G. Parisi, M. Virasoro, "Spin Glass Theory and beyond", Word Scientific, Singapore (1987).
4) See for example D. Ruelle, "Statistical Mechanics", Benjamin (1969).
5) The mean field approach to these problems is described in M. Mezard and G. Parisi, J. Phys. Lett. 46, L771 (1985).

BOOLEAN NETWORKS WHICH LEARN TO COMPUTE

Stefano Patarnello and Paolo Carnevali
IBM ECSEC
Via Giorgione 159
00147 Rome
Italy

ABSTRACT

Through a training procedure based on simulated annealing, Boolean networks can 'learn' to perform specific tasks. As an example, a network implementing a binary adder has been obtained after a training procedure based on a small number of examples of binary addition, thus showing a generalization capability. Depending on problem complexity, network size, and number of examples used in the training, different learning regimes occur. For small networks an exact analysis of the statistical mechanics of the system shows that learning takes place as a phase transition. The 'simplicity' of a problem can be related to its entropy. Simple problems are those that are thermodynamically favored.

The study of the collective behavior of systems of 'formal neurons' which are designed to store a number of patterns ('associative memories') or to perform a task has recently gained increasing interest in physics and engineering applications as well as in biological science [1].

As far as models with biological motivations are concerned, many efforts have clarified, with numerical and analytical methods, the behavior of Hopfield's model [2,3]. Systems with asymmetric 'synapses' which appear to be a more realistic model, have also been proposed [4]. The study of the storage capacity of such systems has taken advantage of methods typical of statistical mechanics, in particular by exploiting the connection between learning systems and spin glasses.

Coming to practical applications in engineering (see [5] and references therein), applications in many areas, including speech synthesis [6], vision [7], and artificial intelligence [8] have been proposed. In these cases less attention has been paid to the general properties of the models, while research has concentrated on the actual capabilities of the systems for specific values of the parameters involved.

In our model [9] we consider networks of N_G boolean gates with two inputs. Each gate implements one of the 16 possible Boolean functions of two variables. Each of its inputs can be connected to another gate in the circuit or to one of the N_I input bits. The last N_O gates produce at their output the N_O desired output bits. To rule out the possibility of feedback we number the gates from 1 to N_G and we do not allow a gate to take input from an higher numbered gate. On the other hand, we ignore fan-out problems allowing each gate to be the input of an arbitrary number of gates. When the gate types and the connections are fixed, the network calculates the N_O output bits as some Boolean function of the N_I input bits.

If we want the network to 'learn' to implement a particular function, we use the following training procedure. We randomly choose N_E examples of values of. the input bits, for which corresponding values of the output bits are known. Then, we try to optimize the circuit in order to minimize the average discrepancy, for these N_E examples, between the correct answer and the one calculated by the circuit. This optimization is performed by simulated annealing [10]: the network is considered as a physical system whose microscopical degrees of freedom are the gate types and the connections. With simulated annealing one then slowly cools down the system until it reaches a zero temperature state, which minimizes the energy. In our case the energy of the system is defined as the average error for the N_E samples.

$$E \equiv \sum_{l=1}^{L} E_l \equiv \sum_{l=1}^{L} \frac{1}{N_E} \sum_{k=1}^{N_E} (E_{lk} - A_{lk})^2 .$$

Here E_{lk} is the exact result from the l-th bit in the k-th example, while A_{lk} is the output for the same bit and example as calculated by the circuit. Therefore A_{lk} is a function of the configuration of the network. Thus, E is the average number of wrong bits for the examples used in the training. For a random network, for example one picked at high temperatures in the annealing procedure, $E_l \sim 1/2$.

As an example, we have considered the problem of addition between two binary integers. We have considered 8-bit operands, so that $N_I = 16$, and ignored overflow (as in standard binary addition), so that $N_O = 8$. In principle the performance evaluation of the system is straightforward: given the optimal circuit obtained after the learning procedure, one checks its correctness over the exhaustive set of the operations, in the specific case all possible additions of 2 L-bit integers, of which there are $N_o = 2^L \cdot 2^L$. This can be afforded for the set of experiments which will be described here, for which $L = 8$ and $N_o = 65536$ Thus another figure of merit is introduced:

$$P \equiv \sum_{l=1}^{L} P_l \equiv \sum_{l=1}^{L} \frac{1}{N_o} \sum_{k=1}^{N_o} (E_{lk} - A_{lk})^2 .$$

This quantity is defined in the same way as E, but the average is done over all possible operations, rather than just over the examples used in the training. We stress that P is only used *after* the training procedure as a tool for performance evaluation.

Roughly speaking, the quantities E and P are all is needed to understand the behavior of the network: low values of E mean that it has been capable at least to 'memorize' the examples shown to it during the training. If P is small as well, then the system has been able to generalize properly since it is able to calculate the correct result for operations it has never been exposed to. Therefore one expects the existence of these two regimes (discrimination and generalization) between which possibly a state of 'confusion' takes place.

A network of 160 gates has been able to organize itself in a completely correct binary adder after a training procedure with $N_E = 224$ only, out of the 65536 possible binary additions of two 8-bit numbers. This means that the system has been able to recognize the rule that was to be used to generate the output, thus generalizing to construct the correct result of any addition not contained in the 224 used during the training. This means that only a fraction .003 of the total samples is necessary to generalize. It is a priori not clear whether or not training could be improved introducing correlations among examples shown, i. e. implementing a sort of 'didactic' teaching.

More generally, we can draw a qualitative picture of the learning processes as they occur in the different cases. As previously mentioned, these are essentially of two kinds. One is lookup-table like: namely, when the system is poorly trained (low N_E), it simply builds a representation of the examples shown, which has nothing to do with any general rule for the operation. Therefore this regime is characterized by values of E near to 0 and values of P near to that of a 'random' circuit, which gives the correct result for each bit with probability 1/2. Therefore $P \sim 1/2 \cdot L = 4$ in this look-up table regime. Providing the system with more and more examples, it will find it hard to follow this brute-force strategy, unless its capability is infinite (the somewhat trivial case $N_G \sim O(N_o)$). Therefore E will increase from 0 as a function of N_E, and P will practically stay constant. As the number of examples used in the training becomes critically high, the onset of the 'generalization regime' occurs provided that the number of gates is large enough, and P will decrease toward 0. This is the region of parameters in which genuine learning takes place.

The specific features for different regimes are somewhat hidden in the 'global' parameters E and P, due to the fact that memorization and learning for each bit start to occur for different N_G and N_E, and are all weakly coupled among each other. Typically the two least significant bits are always correctly processed, and one can roughly say that, as complexity grows when considering more significant binary digits (because of the potentially high number of carry propagations needed), learning 'harder' bits is in a way equivalent to work with less gates. To get a clearer insight in the whole process it is better to focus the attention on the behavior of central bits (to minimize 'border' effects) plotting the quantities E_i and P_i introduced in previous formulae. Figs. 1a, 1b, and 1c are obtained for N_G fixed respectively at 20, 40, and 160. One can recognize the following distinct behaviors:
a) At low N_G (Fig. 1a) only look-up table behaviour occurs. Storing of examples is perfect until $N_E \sim \overline{N}_E = .4 N_G$, which estimates the capacity of the system. It is remarkable that after this value is reached the system does *not* enter a confusion state. In other words this maximum number of 'patterns' is preserved, and simply no more examples are kept. As a consequence, for $N_E > \overline{N}_E$ one has

$$E_l \sim 1/2\,(1 - \frac{\overline{N_E}}{N_E}).$$

In the look-up table region $P_l = 1/2$ for all N_E.

b) For intermediate N_G there is a cross-over to partial generalization. This is clearly shown in Fig. 1b where P_l shows a decrease from $P_l = 1/2$ to a 'residual' value still greater than 0.

c) Finally for large N_G (say $N_G > \overline{N}_G$) the system is able to switch from a perfect storing regime ($E_l = 0$, $P_l = 1/2$) to a complete generalization ($E_l = 0$, $P_l = 0$). For N_G very large we expect this transition to be abrupt, i. e. there is not an intermediate regime where partial generalization takes place. To put it in another way, we conjecture that in this limit there is a critical number of examples N_E^c such that correspondingly the systems switches from perfect storing to complete generalization.

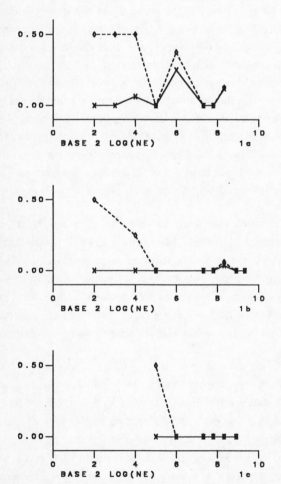

Fig. 1. Behavior of E_l (solid lines) and P_l (dashed lines) as a function of N_E, for various values of N_G (see text).

To summarize this first part, the learning behavior of the system is dependent on its size (N_G), on the complexity of the problem, and on the number of examples used in the training (N_E). For N_G and N_E large enough for the problem to be solvable, generalization and learning take place as described above. If N_G is decreased, the system is no longer able to generalize. For small N_E and for any N_G the system is not able to generalize, but may be able to 'memorize' the N_E examples and construct a circuit that gives the correct answer at least in those N_E cases, or in a significant fraction of them.

Given an explicit example in which the training has led to a network configuration which implements the problem correctly, we want now to address the most puzzling question: how is it that such system is able to perform correctly over *all* possible cases, when given information only on a partial set of examples? In other words, where does generalization come from?

For small enough networks one can study in detail all the properties of the system through a complete enumeration of all possible circuits. As an example, we will refer in the following to a network with $N_G = 4$, $N_I = 4$ and $N_O = 1$. Thus, one can calculate the thermodynamical properties of the system, as well as, for any rule, the average learning probability as a function of N_E and N_G. This analysis entirely confirms the picture sketched above containing the different learning behaviors. In addition, a direct calculation of the specific heat as a function of temperature clearly shows the existence, for most rules, of a peak which, in the limit of large systems, would transform in a singularity characteristic of a phase transition. The intensity of this peak is higher for more 'difficult' rules. Thus, learning clearly appears to be a process of ordering that takes place, when temperature is lowered, in a phase transition. We have been able to recognize a hierarchical structure for the feasible rules, with some degree of ultrametricity.

The analysis based on complete enumeration also clearly indicates that the simplicity of a rule is related to its entropy: simple rules are those that have a large entropy, which means that can be realized in many different ways. As a matter of fact, this kind of approach allowed us to compute *exactly* the learning probability for a given problem, as a function of the number of examples N_E used in the training [11]. This quantity measures the probability that, performing the training with N_E examples, the network will organize in a configuration which implements correctly the problem for *all* possible inputs. In the following we report results on some particular problems.

Let's start by studying the training on a very simple problem, consisting of producing a value of 0 at the output bit regardless of the values of the input bits. In Fig. 2, curve a, we plot the probability of learning as a function of N_E . The curve is for a network with $N_G = 4$. The curve rises quite fast, and reaches 50% already for $N_E = 2$, thus showing that for that N_E the training has 50% probability of resulting in a *perfect* network, i. e., one that produces always 0 at its output, even for the $16 - 2 = 14$ input configurations not used in the training. This already shows clearly the generalization capabilities of the system we are considering. This fast rise of the learning curve is related to the fact that there are very many circuits that always produce zero at their output. In fact 14% of all possible networks with $N_G = 4$ implements the '0 function'.

Now let's consider a more difficult problem, consisting of reproducing at the output bit the value of a specified input bit. The corresponding learning probability is plotted in Fig. 2, curve b, (again the curve is valid for $N_G = 4$). Generalization still occurs, but now we need $N_E = 4$ to get 50% chances of finding a perfect network. At the same time only a fraction $\sim 3.5\%$ of the total number of configurations of the network solve this problem.

We then turn to the even more difficult problem of producing at the output of the network the AND of 3 of the 4 input bits. This problem is solved by a much smaller number of circuits (.047% of the total number). From the plot of the corresponding learning probability (Fig. 2, curve c) one can see that generalization almost does not occur at all, and N_E quite close to 16 (which amounts to give complete information describing the problem to be solved) is needed for the learning probability to be reasonably different from zero ($N_E = 11$ for 50% learning probability). It is clear at this point that the occurrence of generalization and learning of a problem is directly related to the fact that that problem is implemented by many different networks and that this provides also a definition (architectural-dependent) for the complexity of a given problem.

In conclusion, the model we have defined has shown clearly a self-organization capability, when trained on a problem. Moreover, we have been able to provide in this context a clear characterization of generalization processes. We believe that this latter issue could provide some useful hints for other classes of learning machines, as well as for the understanding of learning in biological systems.

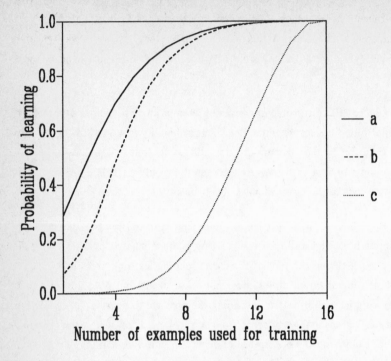

Fig. 2. Learning probability as a function of N_E for three problems.

REFERENCES

1. Hopfield, J.J: Proc. Nat. Acad. Sci. USA, Vol. 79 p. 2254 (1982)
2. Amit, D.J., Gutfreund, H. and H. Sompolinsky: Phys. Rev. A, Vol. 32 p. 1007 (1985)
3. Amit, D.J., Gutfreund, H. and H. Sompolinsky: Phys. Rev. Lett.: Vol. 55 p. 1530 (1985)
4. Parisi, G.: Jour. Phys. A (Math. Gen.), Vol. 19 p. L675 (1986)
5. Personnaz, L., Guyon, I. and G. Dreyfus: in Disordered Systems and Biological Organiza-tion, (Eds. E. Bienenstock et al.), Springer & Verlag (1986)
6. Sejnowsky, T.J. and C. Rosenberg: John Hopkins University Tech. Rep., Vol. 86/01 (1986)
7. Hinton, G.E., and T.J. Sejnowsky: Proc. IEEE Comp. Soc. Conference on Computer Vision and Pattern Recognition, p. 488 (1983)
8. Cruz, C.A., Hanson, W.A. and J.Y. Tam: in Neural Networks for Computing, Am. Inst. of Phys. Proc., Vol. 151 (1986)
9. Patarnello, S. and P. Carnevali: Europhys. Letts., Vol. 4(4) p. 503 (1987)
10. Kirkpatrick, S. Gelatt, S.D. and M.P. Vecchi: Science, Vol. 220, p. 671 (1983)
11. Carnevali, P. and S. Patarnello: Europhys. Letts., in Press

A Dynamical Learning Process for the Recognition of

Correlated Patterns in Symmetric Spin Glass Models

U. KREY and G. PÖPPEL

Institut für Physik III der Universität Regensburg,
D-84 Regensburg, F.R.G.

Abstract. In the framework of spin-glass models with symmetric
(multi-spin) interactions of even order a local dynamical
learning process is studied, by which the energy landscape is
modified systematically in such a way that even strongly correlated noisy patterns can be recognized. Additionally the basins
of attraction of the patterns can be systematically enlarged by
performing the learning process with noisy patterns. After com-
pletion of the learning process the system typically recognizes
for two-spin interactions as many patterns as there are neurons
($p \simeq N^{m-1}$ for m-spin interactions), and for small systems even
more ($p > N$ for m = 2).

The dependence of the learning time on the parameters of the
system (e.g. the average correlation, the noise level, and the
number p of patterns) is studied and it is found that in the
case of random patterns for $p < N$ the learning time increases
with p as p^{x}, with $x \simeq 3.5$, whereas for $p > N$ the increase is
much more drastic.

Finally we give a proof for the convergence of the process and
also discuss the possibility of a drastic improvement of the
learning capacity for patterns with particular correlations
("patched systems").

I. Introduction

Among the most complex processes in nature one should certainly mention the recognition and learning tasks performed by the human brain. Furthermore it seems that the brain recognizes not just by what could be called a systematic programmed search but rather by more indirect associative processes; and early learning, too, apparently happens rather indirectly and slowly through continuous repetition and reinforcement of examples, and not by conscious implementation and derivation of grammatical rules. In fact, the rules - if at all - may be known to the brain only implicitly through associative generalization from the examples.

Therefore, the question should be posed, whether the above-mentioned complex properties of the brain can be simulated in the framework of physical models. Actually this question is a rather old one. Already decades ago models for associative pattern recognition has been suggested. These models have recently been studied intensively by the spin-glass physics community [1]. On the other hand, the problem of learning, to which the present paper is dedicated, has been rediscovered only very recently although already 25 years ago, within the so-called "perceptron concept", basic ideas had already been worked out [2,3].

In the following chapters we define a learning algorithm and study its performance and possible generalizations, by which an associative pattern-recognition system can learn to recognize a considerable number of strongly correlated and very "noisy" patterns. As is well known, this is not possible with the usual Hebb system, see below, which can only recognize up to $p = 0.14 N$ patterns (where N is the number of neurons), and this only if these patterns are uncorrelated, whereas in our algorithm p can be as large as N or even larger (see below).

II. Model description and learning algorithm

We consider the usual Ising spin-glass Hamiltonian

$$H = - \sum_{j,k} J_{jk} S_j S_k \tag{1}$$

where the indices $j, k = 1, \ldots, N$ denumerate the N neurons. The two
states of these neurons ("firing" or "not firing") are represented by the Ising variables $S_j = \pm 1$, and the coupling constants
J_{jk}, which vanish for $j = k$, describe the mutual interactions of
the neurons through synaptic links. Then there are p different
"patterns" $\vec{\xi}^\mu$ out of the 2^N possible spin configurations; these
patterns may be correlated, i.e. the overlap function $q_{\mu\nu} :=$
$(1/N)$ $(\vec{\xi}^\mu, \vec{\xi}^\nu)$, where $(\vec{\xi}^\mu, \vec{\xi}^\nu)$ denotes the usual scalar product of
real N-component vectors, can be different from zero. After the
learning process, by which the J_{jk} are changed (see below),
these patterns should be recognized by the system (1) through the
usual sequential relaxation process [4].

The Hamiltonian (1) can be generalized by taking into account
additional multi-spin interactions of even order. In this way one
is lead to:

$$H = - \sum_{m=2,4,\ldots}^{m_o} \sum_{j_1,\ldots,j_m} J_{j_1,\ldots,j_m} S_{j_1} \cdots S_{j_m} \tag{2}$$

where the couplings are invariant under permutation of the indices and vanish, if all indices are equal (i.e. there is no
constant term). Here it should be noted that in (2) only <u>even</u>
values of m are considered, i.e. the energy H assumes the same
value for a given state $\vec{S} = (S_1, \ldots, S_N)$ and its negative copy
$\underline{\vec{S}} = - \vec{S}$. Thus, if our system recognizes a certain number p of
patterns $\vec{\xi}^\nu$, starting from noisy input state $\vec{\eta}^\nu$, it will usually

also recognize the corresponding negative copy - ξ^{ν}, starting from - $\vec{\eta}^{\nu}$. Actually, however, unless otherwise stated, we always assume in the present paper that the set p of patterns considered contains only <u>one</u> representation of a given pattern and not the corresponding negative copy; then, after the learning process for these p "positive patterns", the system can actually recognize, and usually even distinguish energetically (e.g. after intro- duction of small terms with odd m into (1)), the negative and positive copies, i.e. depending on the personal taste or on possible applications or extensions (see e.g. chapt. IV) the "capacity" of the system may be defined as p or 2p, which should be kept in mind below.

To be specific, our relaxation procedure proceeds as follows: Of three consecutive cycles, during the first and third cycle, N times a position j is selected randomly and the spin S_j is flipped if this leads to a lower energy, while during the second cycle the spins are visited sequentially. Thus we try to avoid that on the one hand some spins are incidentally overlooked, as might be the case by a completely random selection of the spins, while on the other hand we avoid any systematic bias, which might be produced e.g. in the learning process if the spins were always visited in the same order.

The learning process proceeds as follows: We start with the Hebb-Hamiltonian, i.e. with

$$J_{jk} = \frac{1}{N} \sum_{\mu=1}^{p} \xi_j^{\mu} \xi_k^{\mu} \qquad (3)$$

Then, a random permutation $\nu_1, \ldots, \nu_p$ of the p patterns is selected; starting with ν_1, certain <u>input vectors</u> $\vec{\eta}^{\nu_i}$ are sub- mitted, one after the other, i.e. for i = 1 to i = p, to the relaxation process described above. These input vectors can be taken either as the original patterns or as some "noisy" modifi- cations of them. For every i, as many relaxation cycles are performed as are necessary to get the system definitely trapped in a local minimum. This minimum state is the <u>output vector</u> $\vec{x}^{\nu_i}$. If it is different from the original pattern $\vec{\xi}^{\nu_i}$, the Hamiltonian (1) is modified as follows: $J_{jk} \longrightarrow J_{jk} + \Delta J_{jk}$, for all pairs (j,k), with

$$\Delta J_{jk} = \lambda \, (\, \xi_j^{\nu_i} \xi_k^{\nu_i} - x_j^{\nu_i} x_k^{\nu_i}) \tag{4}$$

and the relaxation of the next input vector proceeds with the new Hamiltonian. In (4), λ is a positive number determining the strength of the correction, and implicitly the speed of the learning process (see below). In principle, the parameter λ could also depend on the pattern considered as well as on the pair (j,k), and can also be changed if the learning process is iterated. However, for simplicity we use only a constant λ. Since both $x_j^{\nu_i}$ and the $\xi_j^{\nu_i}$ can only take the values ± 1, the expression (4) can also be written as

$$\Delta J_{jk} = 2 \lambda \, n_{jk}(\nu_i) \, \xi_j^{\nu_i} \xi_k^{\nu_i} \tag{5}$$

where $n_{jk}(\nu_i)$ is either 0 or 1, depending on whether $x_j^{\nu_i} x_k^{\nu_i} = \xi_j^{\nu_i} \xi_k^{\nu_i}$ or not. This means that the learning process leads also to reduction of frustration.

A generalization of (4) and (5) to the Hamiltonian (2) would be:

$$\Delta J_{j_1 \cdots j_m} = \lambda^{(m)} \, (\, \xi_{j_1}^{\nu_i} \cdots \xi_{j_m}^{\nu_i} - x_{j_1}^{\nu_i} \cdots x_{j_m}^{\nu_i}) \tag{6}$$

with $\lambda^{(m)} > 0$. In the following, unless otherwise stated, $\lambda^{(m)}$, which in principle could also depend on ν_i and $j_1, \ldots, j_m$, is assumed to be constant for $m = 2, 4, \ldots, m_0$ and to vanish for $m > m_0$.

The teaching process is stopped, if for all members of the permutation considered the output vectors $\vec{x}^{\nu i}$ are identical with the patterns $\vec{\xi}^{\nu i}$ to be learned (of course this condition can be weakend or modified); otherwise it is repeated with a new permutation and with new noisy modifications of the corresponding patterns. Of course, one always uses the latest version of the corrected Hamiltonian.

Our numerical analysis have been performed for $m_0 = 2$. In general, a very large number of iterations is necessary for successful learning particulary if p is comparable to N (see below), however in practice we have always found that after a sufficent number of iterations the learning process stopped. In fact, in the appendix we give a proof that the learning stops almost certainly, at least if it is performed with the pure patterns.

In any case, one can easily interprete the physics behind our procedure by considering the energy change

$$\Delta H(\vec{S}) = -\sum_m \sum_{j_1 \cdots j_m} \Delta J_{j_1,\ldots,j_m} \, S_{j_1} \cdots S_{j_m}$$

$$= \sum_m \lambda^{(m)} \, [\, (\vec{x}^{\nu i}, \vec{S})^m - (\vec{\xi}^{\nu i}, \vec{S})^m \,]$$

(7)

induced by the correction (6) for a given spin configuration $\vec{S}$. For $\vec{S} = \vec{x}^{\nu i}$, i.e. for the "unwanted" output resulting from the input vector corresponding to pattern $\vec{\xi}^{\nu i}$ (e.g. a noisy modification), ΔH is positive, whereas for the pattern $\vec{\xi}^{\nu i}$ it is negative, namely

$$\Delta H(\vec{x}^{\nu i}) = -\Delta H(\vec{\xi}^{\nu i}) = \sum_m \lambda^{(m)} N^m \{\, 1 - [\, q(\vec{x}^{\nu i}, \vec{\xi}^{\nu i})\,]^m \,\} \qquad (8)$$

Thus, the essential point of our learning process consists in a systematic increase (decrease) of the energy of unwanted (wanted) states, i.e. the energy landscape in the vicinity of the patterns $\xi^{\nu i}$ is modelled in such a way that these patterns become local minima. Moreover, by performing the learning process not with the original patterns, but with noisy modifications of them, one can systematically enlarge the basins of attraction of the patterns. At the same time, the probability to have unwanted (i.e. "spurious") minima should be strongly reduced.

III. Results

In our numerical studies we concentrate on the usual case $m_0 = 2$ and define $\lambda = \lambda^{(2)}$ N. In Fig. 1 results are presented, which characterize the improvement of the recognition process through the learning procedure for a system with N = 100 neurons and p = 10 patterns with an averaged correlation of $\bar{q}_{\mu\nu}$ = 0.118 ± 0.087:

The <u>retrieval quality</u>, i.e. the overlap $q(\vec{\xi}^1, \vec{x}^1)$ of the original pattern $\vec{\xi}^1$ and the stationary output $\vec{x}^1$ of the relaxation is plotted over the noise level p_n (fraction of randomly flipped spins). Every point on the curves of Fig. 1 represents an average over 100 different noisy modifications of pattern $\vec{\xi}^1$ with identical noise level, and the error bars represent the standard deviation from the average. (Actually the distribution of the 100 results for a given p_n is strongly non-Gaussian, with a sharp peak at the maximum value of $q(\vec{\xi}^1, \vec{x}^1)$ in each curve, i.e. at 0.55 in Fig. 1a, and at 1 in Fig. 1b, and a broader distribution centered around a smaller value, appearing for noise levels > 0.3).

In any case, from a comparison of Fig. 1a with Fig. 1b it is obvious that the retrieval quality has been drastically improved through the learning process (Fig. 1b), i.e. after a learning process which has been performed with <u>noisy</u> patterns, with a noise level $\sigma = 0.3$.

If the learning process would be performed with pure patterns ($\sigma = 0$), the retrieval quality would still be much better than

in Fig. 1a, but not as good as in Fig. 1b: The reason is that for
σ = 0.3, in contrast to σ = 0, there is an additional enlargement
of the basins of attraction by the learning process.

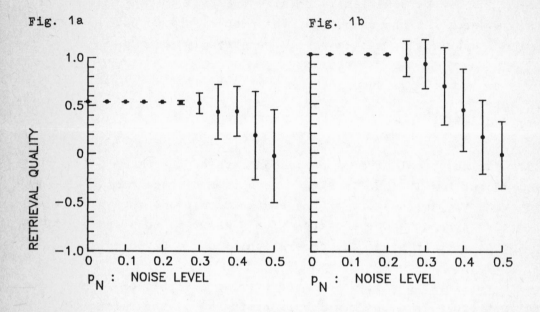

Fig. 1a

Fig. 1b

Fig.1 The retrieval quality, i.e. is the averaged overlap
$q(\vec{\xi}^1, \vec{x}^1)$ of the original pattern $\vec{\xi}^1$ and the output $\vec{x}^1$ of the
relaxation process, is presented for 100 relaxation processes
starting with different noisy modifications of the original pattern, as a function of the noise level p_n, i.e. the relative
number of randomly flipped spins of the modifications, for a
system with N = 100 neurons and p = 10 patterns, which have an
averaged correlation of $\bar{q}_{\mu\nu}$ = 0.118 ± 0.087. **Fig. 1a** is for the
Hebb-system, Eq. (3), i.e. without learning, **Fig. 1b** for the
system as obtained after a learning process where additionally
the basins of attraction have been enlarged by using noisy input
patterns with a noise level of σ = 0.35. The learning strength
was λ = 0.0005, and the total number R of learning steps was
5222 in case b.

Furthermore, using a sufficiently long teaching time, we have found that for small systems 100% retrieval can be obtained even for <u>more</u> than N patterns, i.e. for p = 60 random patterns in case of N = 50, see below.

Of course, the necessary number R of <u>teaching steps</u> (i.e. corrections of the Hamiltonian) increases strongly with p, α, $\bar{q}_{\mu\nu}$ and $1/\lambda$.

We have found that R is proportional to $1/\lambda$ and proportional to $\bar{q}_{\mu\nu}$ at least for $0.01 \leqslant \lambda \leqslant 0.08$ and $0.10 \leqslant \bar{q}_{\mu\nu} \leqslant 0.20$, with $R \simeq 400$ for $\bar{q}_{\mu\nu} = 0.1$ and $\lambda = 0.01$ (with N = 100, p = 20).

The dependence on the number of patterns can be seen from Fig. 2a. For systems with N = 50, 100 or 200 neurons the number R of learning steps for correct recognition of p uncorrelated patterns increases strongly with p: However, for p < N the increase seems to be non-exponential, e.g. for N/2 < p < N, R is found to behave as $\sim p^x$, with $x \approx 3.2$ for N = 100 and ≈ 3.6 for N = 50. Only for $p \geqslant N$, R increases more drastically; however from our data we cannot make a definite statement whether e.g. for N = 50 and $50 \leqslant p \leqslant 60$ the increase is $\sim p^y$ with $y \approx 9$, or whether the increase is even exponentially. In fact, for a generic spin glass, i.e. with a Gaussian exchange, one has <u>exponentially</u> many local minima [5] and would of course expect an exponentially large teaching time. Thus one may speculate that the models prepared by our teaching process may somehow interpolate between the separable Hebb model and the generic case, and that the crossover happens around $\alpha \simeq 1$.

Finally, in Fig. 2b, the dependence of the learning time on the averaged correlation $\bar{q}_{\mu\nu}$ is studied. Again, the increase is not very drastic, except around extremely large values, i.e. for $\bar{q}_{\mu\nu} > 0.6$.

Fig. 2a

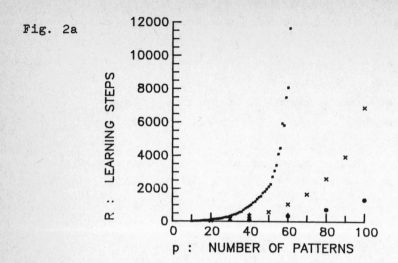

Fig. 2b

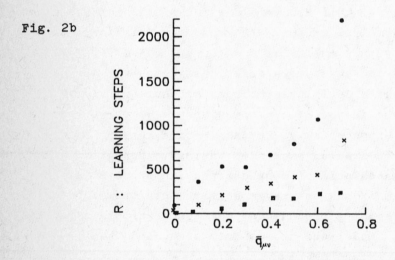

Fig. 2 The number of learning steps R is studied as a function of various system parameters, namely in Fig. 2a as a function of p up to very large values of p, for $\lambda = 0.04$ and $\bar{q}_{\mu\nu} \approx 0$ (circles: N = 200, crosses: N = 100, squares: N = 50); - in Fig. 2b as a function of $\bar{q}_{\mu\nu}$ up to very large correlations, again with $\lambda = 0.04$ (circles: N = 200, p = 40; crosses: N = 100, p = 20; squares: N = 50, p = 10) .

As already mentioned, for the enlargement of the basins of attraction it is necessary to perform the learning process with noisy modifications of the original patterns. In Fig. 3 we study the question, whether the corresponding noise level σ prolongs the learning time. As can be seen from Fig. 3, for $\sigma < 0.2$ this is practically <u>not</u> the case, and also for the very large value $\sigma = 0.3$, the increase of R compared with $\sigma < 0.2$ amounts only to a factor 2.

Fig. 3

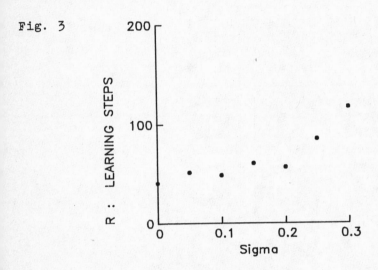

Fig. 3 The total number of learning steps R for the learning process with <u>noisy</u> patterns (cf. Fig. 1b) is presented over the noise level σ. N = 100, p = 10, λ = 0.02, $\bar{q}_{\mu\nu}$ = 0.118 ± 0.087.

To make the retrieval quality and the possible applications obvious, we present in the final Fig. 4 a recognition process, where in a system with N = 256 neurons p = 6 extremely correlated patterns, namely the letters A,B,C,D,E,F, which have an averaged correlation of 0.78, are recognized after just three relaxation cycles, although these patterns are presented to the system "in strong disguise" (see the second column), corresponding to a noise level of p_n = 0.3, so that the human eye would no longer

recognize them. The number of learning steps, which were performed with $\lambda = 0.05$ and $\sigma = 0.3$, was R = 333. Of course, still
stronger "disguise" of the patterns would hardly make sense in
the present example. In any case at this place we would like to
stress that it is not primarily the Hamming distance, but rather
the sculpturing of the energy landscape, i.e. our learning process, which determines whether e.g. the first noisy pattern is
recognized as a "A" and not as a "B".

Fig. 4

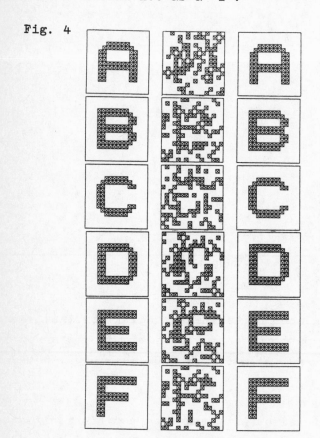

Fig. 4 Pattern recognition of the letters A,B,C,D,E,F, as explained in the text. The first column represents the pure patterns, the second column the noisy modifications, which are taken
as the input vectors of the recognition process, and the third
column the output of the recognition, which took three relaxation
cycles. (Learning parameters: $\lambda = 0.05$, $\sigma = 0.3$, R = 333)

IV. Remarks on patched systems

In principle, the capacity of $\alpha = p/N \approx 1$ achieved by our learning process in Fig. 2b, although much larger than that of the Hebb prescription, is quite small, when compared with the total number 2^N of states of the system. However, one should note that this capacity was found numerically for random patterns and may be improved for correlated patterns by taking advantage of the correlations. This can most clearly be seen by considering the following case of hierarchically correlated patterns in a system with N neurons: We assume, (I), that the system is subdivided into q separate "patches" with N/q neurons; then, (II), within each patch (j) a given set of "small patterns" $\vec{\xi}_\nu^{(j)}$, $\nu = 1,\ldots,p$, is defined, and finally for the global systems the set of patterns to be learned is defined to consist of all possible combinations of the small patterns, i.e.

$$\vec{\xi}_\nu = (\vec{\xi}_{\nu_1}^{(1)}, \vec{\xi}_{\nu_2}^{(2)}, \ldots, \vec{\xi}_{\nu_q}^{(q)}) \qquad (9)$$

Now, since within each layer at least up to $2p \lesssim 2\,N/q$ small patterns can be learned (if the negative copies are counted, too) the total system can actually recognize all $(2p)^q$ ($\lesssim (2\,N/q)^q$) combinations $\vec{\xi}_\nu$, i.e. $1/2\,(2p)^q$ <u>globally</u> positive copies, a number which can be much larger than the number N of neurons. To achieve this drastically enhanced performace one only has to make all J_{jk}- parameters vanish, which join neurons from different patches; then within each layer the learning process can be performed simultaneously. Thus, in this particular case, correlation leads to an exponential enhancement of performance at almost no price.

Generally, of course, the situation is more complex: Not every combination of small patterns (e.g. letters) will be equally probable, e.g. not every combination of letters yields a meaningful expression in the context of a given language. However one may introduce a suitable generalization of the model (2), where now the indices (j) represent "patches" or "layers" and the spins S_j assume as many values as there are "letters". For this system the global patterns will be "meaningful words", which can hopefully be learned analogously to the approach described in the preceding chapters. However, one might also consider directly the global problem right from the start to see whether our learning algorithm itself will <u>implicitly</u> discover the existing correlations and use them optimally, without being guided too much from outside. This question is presently under numerical study, and not much is known on the answer at present. However, in any case, there is hope that the answer may be positive, since at least in the simple case of the "patched" correlations discussed above, our convergence proof, which is given in the appendix, works even if the connections between different patches are <u>not</u> forced to vanish right from the start.

Possibly also the fact that for N = 50 our system could learn at least p = 60 patterns chosen at random, (see chapt. III), may be due to the "residual correlations" which must be present in this system, leading to roughly $\approx \sqrt{50} \approx 7$ additional patterns, which can be learned beyond p = N. Also Gardner et al. have found a drastically enhanced capacity for a particular correlated system.

V. Conclusion

We have defined a learning process for the associative recognition of strongly correlated patterns within the framework of spin-glass models. The learning algorithm is based on a modelling of the energy landscape by which the "wanted" states (i.e. the patterns) are lowered in energy and the "unwanted" states enhanced. Furthermore, by performing the learning process with noisy patterns, one can systematically enlarge the basins of attraction of the patterns. Similar learning algorithms, which differ only in detail, but not in spirit, have been independently suggested

and studied recently by a number of the other authors [6,7,8].
Actually all these algorithms seem to belong to the general class
of "perceptron learning algorithms" studied already 25 years ago
([2,3], see also D.J. Amit, these proceedings). Compared with
the approach of Gardner et al. [6] and Diederich and Opper [7]
the present algorithm seems to be slightly more "natural" and
implicit, since in our case the full relaxation to the next
metastable state is performed before the J_{jk} are corrected, and
since our correction, see eq. (6), does not need an error mask.
Furthermore, our learning process may be made still more implicit, if in the correction prescription (6) the original patterns
$\vec{\xi}^{\nu}$ would be replaced by suitable time averages of noisy input
patterns, which would converge to the "pure" patterns only in
course of the learning process: This would represent some kind of
"emergence of meaning through repetition and association".

We have discussed the performance of the learning algorithm and
found that in the simplest case of two-spin interactions ($m_0 = 2$)
at least $p \simeq N$ patterns and for m_0-spin interactions at least $p \simeq$
N^{m_0} patterns can be learned. Moreover, as discussed in chapt. IV,
for certain hierarchically correlated patterns even with two-spin
interactions one may obtain capacity p as large as $(2N/q)^q$ with
$1 << q << N$, i.e. p would be much larger than the total number N
of neurons. In the appendix it is shown that our dynamical learning process works even in that case, without explicit knowledge
of the correlations.

Concerning the number R of necessary learning steps for random
patterns it was found for $m_0 = 2$ that R increased as p^x, with
$x \approx 3.5$ as long as $p \lesssim N$, with a drastic increase beyond $p \approx N$,
while at the same time R did hardly depend on the noise level of
the "noisy input patterns" except for extremely large values.

Finally we would like to remark that our "learning neuron
network" may be termed complex, since it works with random processes and noise, and since also the success of a learning process is not completely predicatble, as in nature.

Appendix

Here for $\lambda^{(m)} = \lambda$ for $m=2,4,\ldots,m_0$, $= 0$ else, we give a rather general proof that our learning process will stop after a finite number of steps under certain conditions, (i), (ii), which will be stated below.

We start with the general multi-spin Hamiltonian (2) and define a vector $\vec{J}$ whose components are the $J_{j_1,j_2,\ldots,j_m}$ - elements, for all possible index - combinations with $m = 2,4,\ldots m_0$. The corresponding changes at a certain learning step ν_i are

$$\Delta J_{j_1\ldots j_m} = \lambda \; (\xi_{j_1}^{\nu_i} \ldots \xi_{j_m}^{\nu_i} - x_{j_1}^{\nu_i} \ldots x_{j_m}^{\nu_i}) \tag{A1}$$

Then we assume, (i), the existence of a certain vector $\vec{J}^*$, for which the energy functional (2) with $\vec{J} = \vec{J}^*$ is not constant and with a global minimum value E_0, which is obtained for all the patterns $\vec{\xi}^\nu$ ($\nu = 1,\ldots,p$) to be learned, but also - maybe - for certain additional states. For $m_0 = 2$ such a set $\vec{J}^*$ can <u>always</u> be explicitly constructed by the pseudo-inverse method of Personnaz et al. [9], at least for p ($< N$) linearly independent patterns. Moreover for correlated patterns $\vec{J}^*$ may even exist for $p \gg N$, e.g. for the particular "patched" set of patterns described in chapt. IV it exists for $p < 1/2 \, (2N/q)^q$, where $q > 1$ is the number of patches. For $m_0 \geq 4$, even for random patterns, p may be as large as

$$\sum_{j=1}^{m_0-1} \binom{N}{j} \approx N^{m_0 - 1} \; (1 + O(\frac{1}{N})) \; ,$$

see [10].

Now, the idea of the proof is to show that with increasing learning time t the scalar product $\vec{J}^* . \vec{J}$ increases $\sim t$, whereas

$|\vec{J}^*||\vec{J}|$ remains smaller than $\sim t^{1/2}$, therefore the Schwartz -
inequality $\vec{J}^*.\vec{J}^* / |\vec{J}^*||\vec{J}| \leqslant 1$ implies that the process must
stop.

At first we consider the change $\Delta (\vec{J}^*.\vec{J}) = \vec{J}^*.\Delta\vec{J}$ of the nomina-
tor by a learning step involving ξ^ν. Equ. (A1) yields

$$\Delta (\vec{J}^*.\vec{J}(t)) = \lambda [H^*(\vec{x}^\nu) - H^*(\vec{\xi}^\nu)] \qquad (A2)$$

Due to (i), (A2) is $\geqslant 0$. If it vanishes, then either $\vec{x}^\nu = \vec{\xi}^\nu$
(= event (a), successful recognition), or $\vec{x}^\nu$ would incidentically agree with one of the other patterns $\vec{\xi}^\mu$ (= event (b)), or
with one of the other states, for wich $H^*(\vec{S}) = E_o$ (= event (c)).

According to our numerical experience, beginning with the Hebb-
matrix, the events (b) and (c) seem to be very rare, which is not
astonishing, since under the assumptions on p mentioned after (i)
the set of states $\vec{S}$ with $H^*(\vec{S}) = E_o$ is for $N \to \infty$ of measure zero
[11,12,13]. Moreover one can additionally reduce the probability
for the events (b) and (c) by replacing (A1) by the weighted
average

$$\Delta J_{j_1\cdots j_m} = \lambda \sum_{i=1}^{f_\nu} p_i (\xi^\nu_{j_1} \cdots \xi^\nu_{j_m} - x^{(\nu,i)}_{j_1} \cdots x^{(\nu,i)}_{j_m}) \qquad (A1')$$

with $p_i > 0$ and $\sum_i p_i = 1$, where $\vec{x}^{(\nu,i)}$ $(i=1,2,\ldots f_\nu)$ is the
sequence of states assumed during the relaxation from $\vec{\eta}^\nu$ to
$\vec{x}^{(\nu,f)} \equiv \vec{x}^{(\nu)}$. With (A1'), (A2) is replaced by

$$\Delta (\vec{J}^*.\vec{J}(t)) = \lambda \sum_{i=1}^{f_\nu} p_i [H^*(\vec{x}^{(\nu,i)}) - H^*(\vec{\xi}^\nu)] \qquad (A2')$$

which is again $\geqslant 0$; however, now the undesired events (b) or (c)
would only happen if incidentically <u>all</u> the $\vec{x}^{(\nu,i)}$ would belong
to the set with $H^* = E_o$.

Now let δ be the smallest positive value which $\vec{J}^* . \Delta\vec{J}$ can assume among the finite number of all states $\vec{S}$; according to (i) and (A2) or (A2') such a value exists. Then, if the events (b) or (c) would never happen, $\vec{J}^* . \vec{J}$ will increase with t at least as $(\delta/p).t$, since for at least one of the p trials of a learning cycle the recognition is unsuccessful, i.e. $\vec{J}^* . \Delta\vec{J} > \delta$.

However, for the following it suffices to assume that, (ii), events (b) or (c), although not necessarily rare, are rare enough that $\vec{J}^* . \vec{J}$ still increases at least as $(\delta/p).c.t^x$, with c > 0 and $1/2 < x \leq 1$.

Then, at least if the learning is performed with the pure patterns $\vec{\xi}^\nu$ instead of noisy ones (i.e. for $\vec{\eta}^\nu = \vec{\xi}^\nu$), the proof would still work:

Now the denominator $|\vec{J}^*||\vec{J}(t)|$ appearing in the Schwartz - inequality is considered. We start from $\Delta\{|\vec{J}(t)|^2\} = 2\,\vec{J}(t).\Delta\vec{J} + (\Delta\vec{J})^2$. Here the first expression on the r.h.s. can be shown to yield

$$2\,\vec{J}(t).\Delta\vec{J} = -2\lambda \sum_{i=1}^{f_\nu} p_i\,[\,H_t(\vec{\xi}^\nu) - H_t(\vec{x}^{(\nu,i)})\,] \tag{A3}$$

where H_t is the Hamiltonian corresponding to $\vec{J}(t)$, i.e. just before the recognition trial considered. If the input vector $\vec{\eta}^\nu = \vec{\xi}^\nu$, then (A3) ≤ 0, since no relaxation step can lead to a higher energy. However, the components of $\Delta\vec{J}$ are smaller than 2λ in magnitude (see eq. (A1), (A1')) ; there $\Delta\vec{J}.\Delta\vec{J} \leq 4\lambda^2 N^m$, where N^m is the number of components of $\vec{J}$. Therefore $\vec{J}(t) \leq 2\lambda N^{m/2}.t^{1/2}$, as stated above. Thus fore pure input patterns the proof is complete.

What about the remaining case of noisy input patterns ? Then, depending on the noise level and on the starting state $\vec{J}(t=0)$ it may often happen that (A3) becomes positive, e.g. $H_t(\vec{\xi}^\nu) < H_t(\vec{x}^\nu) < H_t(\vec{\eta}^\nu)$. If this does not happen too often, i.e. if $\vec{J}(t).\Delta\vec{J}$ remains $\leq d.t^y$, with $0 \leq y < 2x-1$, then $\vec{J}(t)^2$ remains smaller than $d'.t^{y+1}$, and $\vec{J}^* . \vec{J}\,/\,|\vec{J}^*||\vec{J}| > c'.t^{(x-y/2-1/2)}$ i.e. the proof would still work.

References

[1] See e.g. the papers of AMIT D. J.; SOMPOLINSKY H.;
     KINZEL W.; HERTZ J. A., GRINSTEIN G. and SOLLA S. A. ;
     VAN HEMMEN J. L.; TOULOUSE G. in : *Heidelberg Colloquium on
     Glassy Dynamics*" ; VAN HEMMEN J. L. and MORGENSTERN I., eds;
     Lecture Notes in Physics, Vol. 275; Springer Verlag,
     Heidelberg 1987.

[2] BLOCK H. D., *Rev. of mod. Phys.* 34 (1962), 123.

[3] MINSKY M. L. and PAPERT S., *Perceptrons*, MIT Press (1969).

[4] BINDER, K. (Editor), *Monte Carlo Methods in Statistical
     Physics*, 2nd Edition, Berlin-Heidelberg-New York :
     Springer Verlag (1986).

[5] BRAY A. and MOORE M.A., *J. Phys. C*, 13 (1980), L-469.

[6] GARDNER E., STROUD N. and WALLACE D. J.,
     Edinburgh preprint 87/394 (submitted to *Phys. Rev. Lett.*).

[7] DIEDERICH S. and OPPER M., *Phys. Rev. Lett.* 58 (1987) 949.

[8] KRAUTH W. and MEZARD M., *J. Phys. A*, 20 (1987) L-745

[9] PERSONNAZ L., GUYON I., and DREYFUS G., *J. Phys. (Paris)
     Lett.*, 46 (1985) L-359.

[10] BALDI P. and VENKATESH S. S., *Phys. Rev. Lett.* 58 (1987)
     913.

[11] KOMLOS J., *Studia Scientiarum Mathematicum Hungarica 2*
     (1967) 7.

[12] VENKATESH S. S. and PSALTIS D., Linear and logarithmic
     capacities in associative neural networks, preprint
     IEEE:IT Rev. 4/24/87.

[13] KANTER I. and SOMPOLINKSKY H., *Phys. Rev. A*, 35 (1987)
     380.

NEURAL NETWORKS THAT LEARN TEMPORAL SEQUENCES

Jean-Pierre Nadal

Groupe de Physique des Solides de l'Ecole Normale Supérieure

24 rue Lhomond, 75005 Paris (France)

Introduction

Networks of formal neurons have been studied intensively, especially since 1982, within the framework of statistical physics [1,2,3]. Models for distributed, content addressable, memories have attracted much attention. Recently it has also been shown how to process temporal sequences in similar networks [4-13]. Most of these works have been devoted to the storage and retrieval of "simple" sequences, that is of sequences made of a set of distinct patterns. In the following I will discuss more especially the storage and retrieval of "complex" sequences, as explained below. I will use this term "complex" essentially with the same meaning as used by P. Grassberger in this meeting - who, however, deals with much more complex sequences than I do ! I want to show that the complexity of a sequence is here naturally related to some complexity in the architecture of the network which can process it. Following H. Atlan, I will also put the emphasis on the differences between biological and ingineering oriented approaches, even though they share the same basic ideas.

Pattern recognition

Networks of formal neurons are made of a large number N of spin like units, S_i, i = 1,N, which can be + 1 (neuron active) or - 1 (neuron quiscent). Each neuron i is connected (possibly) to every other neuron j, and the synaptic efficacy J_{ij} is the strength of the interaction. Starting from some initial condition $S_i(t = 0)$, the network evolves under the dynamics

$$S_i(t + \Delta t) = \begin{cases} + 1 \text{ with probability } 1/\left(1 + \exp - 2\beta h_i(t)\right) \\ - 1 \text{ with probability } 1/\left(1 + \exp + 2\beta h_i(t)\right) \end{cases} \quad (1)$$

where β^{-1} is a temperature like parameter and

$$h_i(t) = \sum_j J_{ij}^s S_j(t) \quad (2)$$

For sequential asynchronous dynamics, $\Delta t = 1/N$, and for parallel dynamics, $\Delta t = 1$. The set of attractors of the system constitutes the set of "memories". For pattern recognition, learning means to choose the J_{ij} as a function of the p patterns $\{(\xi_i^\mu, i = 1, N), \mu = 1, p\}$, so that these patterns, or at least configurations very similar to them, are memories of the network. This means, in particular, that without noise ($\beta^{-1} = 0$), each pattern μ has to be (meta)stable, that is

$$\sum_j J_{ij}^s \, \eta_{ij}^\mu \geq K \tag{3}$$

for some $K \geq 0$, with

$$\eta_{ij}^\mu = \xi_i^\mu \xi_j^\mu \tag{4}$$

and this for all i if exact retrieval is required, for most i otherwise. The superscript s in J_{ij}^s refers to the *stabilization* role of these weights.

In the "standard" model [1], the learning scheme follows the empirical Hebbian rule :

$$J_{ij}^s = \frac{1}{N} \sum_{\mu=1}^{p} \xi_i^\mu \xi_j^\mu \tag{5}$$

If $\alpha = p/N$ is smaller than a critical value $\alpha_c \cong 0.14$, for each pattern there is a memory similar at 97% at least to that pattern [1,3]. This model is one of the simplest which allows for a detail study with the tools of statistical physics. For practical applications, however, one has to chose some more efficient rule. It has been shown that, if one asks for exact retrieval, the maximal possible capacity is $\alpha_c = 2$ [14,15]. Several iterative algorithms have been proposed [15-17], which allows to reach such a high capacity. They are essentially variant of the "Perceptron algorithm" [18] (see also D.J. Amit, this meeting), solving equations (3). An other type of choice consists in replacing (3) by the sufficient condition

$$\text{for all } i, \mu \quad \sum_j J_{ij}^s \, \xi_j^\mu = \xi_i^\mu \tag{6}$$

This linear system of equations can be solved explicitly [19,20]. The solution is a function of the pseudo-inverse [21] of the matrix whose p columns are the components of the patterns. Here the critical value for linearly independent patterns is $\alpha_c = 1$.

Processing temporal sequences

Now we would like to associate the p patterns in a sequence, so that, giving as initial condition a configuration identical or very similar to the first pattern ξ^1, the network evolves, going successively from one state ξ^μ to the next one $\xi^{\mu+1}$. At this point there is an important difference between artificial and biological applications. In the first case, paralell dynamics is best suited for fast computation, and since the meaning of each state ξ^μ is known by the user, it is sufficient that the system evolves at each time step from one pattern to the next one. In the second case, however, sequential asynchronous updating is more realistic. More importantly, there should be a way such that the system can realize that it is in a given state ξ^μ. For patterns recognition, this is obtained by the stabilization of the activity in a (meta-)stable state. Thus, here one ask for the system to remain during some time in the configuration μ, and then to make a sharp transition towards the next pattern $\mu + 1$.

Consider first the ingineering approach, which is simpler. We have just to modify the constraints (3), (4) in

$$\sum_j J_{ij}^t \, \eta_{ij}^\mu \geq K \tag{7}$$

$$\eta_{ij}^\mu = \xi_i^{\mu+1} \, \xi_j^\mu \tag{8}$$

Indeed if (7) is true, if at time t $S_i = \xi_i^\mu$, then the local field h_i is of the sign of $\xi_i^{\mu+1}$. As in (6), a particular solution is obtained by the choice :

$$\text{for all i}, \mu \quad \sum_j J_{ij}^t \, \xi_j^\mu = \xi_i^{\mu+1} \tag{9}$$

Here the superscript t stands for *transition* synaptic weights. To solve (7), (8), the Perceptron type algorithms quoted above can be used [7], since they rely only on the set of values (+ or - 1) of the η_{ij}^μ, and the linear system (9) can be solved again with the pseudo-inverse technics (under some condition of solvability) [7].

Now we know how to build a matrix J_{ij}^s which stabilizise a pattern, and a matrix J_{ij}^t which provoque the transitions $\mu \to \mu+1$. How can we put these together to deal with the biologically oriented approach ? The simplest way would be to add the two contributions:

$$J_{ij} = J_{ij}^s + J_{ij}^t \tag{10}$$

This, however, is not efficient, as noted by Hopfield [1], since the patterns become rapidly mixed. Recently a rule, similar to (10), has been proposed [9]. However its efficiency seems to rely on the choice of patterns which do not overlapp. The simplest and still biologically plausible idea, as first shown by P. Peretto and J. Niez in 1985 [4], is to consider two type of synapses, having different relaxation times. A simplification, which do not affect the qualitative results, is to assume a delay time τ, so that (2) becomes

$$h_i(t) = \sum_j J_{ij}^s \, S_j(t) + \sum_j J_{ij}^t \, S_j(t-\tau) \tag{11}$$

The model defined by (11), with the Hebbian rules

$$J_{ij}^s = \frac{1}{N} \sum_\mu \xi_i^\mu \, \xi_j^\mu \quad , \quad J_{ij}^t = \frac{\lambda}{N} \sum_\mu \xi_i^{\mu+1} \, \xi_j^\mu \tag{12}$$

(where λ is a parameter measuring the relative strength of the transitions term to the stabiliting term), has been studied on the fully connected network [5] and on a strongly diluted network [11]. In this latter case, the dynamics can be solved exactly in the large τ limit. The results then obtained examplifies the effect of noise. The noise has two contributions : one is the intrinsic noise, measured by β, the other one is due to the superposition of many patterns, and depends on the two parameters λ and $\alpha = p/N$. In particular, noise prevents the system from being trapped in spurious states. Also, at a given value of λ, a sufficiently large α is required so that a sequence can be retrieved. The critical capacity of the network is maximal at $\lambda = 1$, and is twice the capacity for pattern recognition. This comes simply from the fact that there is twice more synaptic efficacies ! In fact, finding optimal coupling, in the large τ limit, means to solves :

$$\text{for all } i \text{ and } \mu \quad \sum_j J_{ij}^s \, \xi_i^\mu \, \xi_j^\mu + \sum_j J_{ij}^t \, \xi_i^\mu \, \xi_j^{\mu-1} \geq K \tag{13}$$

Choosing the normalizations :

$$\sum_j (J_{ij}^s)^2 = N \quad , \quad \sum_j (J_{ij}^t)^2 = \lambda^2 N \tag{14}$$

one can compute the maximal possible capacity [22], in the very same way as for pattern recognition. If we call $\alpha_p(K)$ the maximal possible capacity for pattern storage computed in [15], then one finds [22] the capacity $\alpha_s(\lambda,K)$ for sequence storage :

$$\alpha_s(\lambda,K) = \frac{(1+\lambda)^2}{1+\lambda^2} \, \alpha_p \left(\frac{K}{\sqrt{1+\lambda^2}} \right) \tag{15}$$

This gives in particular $\alpha_c(1,0) = 2\,\alpha_p(0) = 4$.

Complex sequences

All what has been said in the previous section is correct provided no two patterns in the stored sequences are identical. Indeed, information on the next state is contained in J_{ij}^t and depends only on the actual state. In a sequence such as $\{1,2,3,2,4\}$, the knowledge of being in state 2 is not enough to make the decision of a transition. 2 is a "bifurcation point" for this sequence. One can define the degree, or order, of a sequence, or of a set of sequences, by the minimal memory span one has to keep in order to produce the sequence [6,12]. A simple sequence is of order 0 - note that a cycle made of distinct patterns is of order 0 -. The set of the two following sequences is of order 1 : ($\{1,2,3,4\}$, $\{5,3,6,2,7\}$). Learning such complex sequences has been first discussed in a biological context [6], and then for engineering applications [12,13]. The storage and retrieval of complex sequences in the context of artificial applications is a simple generalization of the storage of simple sequences [7]. For simplicity I will consider here only the case of sequences of order 1. In this case we have to take into account possible bifurcation points, for which the updating (1) at time t must depends on the activities at times t and t - 1. Hence (2) has to be modified [12] :

$$h_i(t) = \sum_j J^0_{ij} S_j(t) + \sum_j J^1_{ij} S_j(t-1) + \sum_{jk} J_{ijk} S_j(t) S_k(t-1)$$

It is sufficient in fact to keep only the linear terms - or to keep only the quadratic term -. Thus I consider the choice

$$h_i(t) = \sum_j J^0_{ij} S_j(t) + \sum_j J^1_{ij} S_j(t-1) \tag{16}$$

If one defines $\gamma(t)$ as the vector of 2N components obtained by concatenation of S(t) and S(t - 1), (16) can be rewritten

$$h_i(t) = \sum_{l=1}^{2N} C_{il} \gamma_l(t) \tag{17}$$

and we have to solve

$$\text{for all } i, \mu \qquad \sum_{l=1,2N} C_{il} \eta^\mu_{il} \geq K$$

where here

$$\eta^\mu_{il} = \xi^{\mu+1}_i \gamma^\mu_l \tag{19}$$

Thus the formulation is strictly identical to the one for simple sequences, with an effective network of 2N neurons. In particular all the algorithms known for choosing the coupling in the case of pattern storage can be, again, used here [12].

For sequential asynchronous updating, one could try to apply the same method, with the generalization of (11) :

$$h_i(t) = \sum_j J^s_{ij} S_j(t) + \sum_j J^{t0}_{ij} S_j(t-\tau) + \sum_j J^{t1}_{ij} S_j(t-2\tau) \tag{20}$$

However this would correspond to synapses of three different relaxation times, and sequences of even higher complexity would require even more different relaxation times ! In the following I present shortly a model which avoid this problem and go further towards biological plausibility.

Listening to the birds

In all the model presented above, the synaptic efficacies are prescribed once we know what has to be learned : there is no true learning. A biological network is a dynamical system, and learning occurs through interactions between the external world and the internal activity. The first aim of the following model [6] is precisely to propose a plausible learning procedure. It takes its justification in data on acquisition of song in birds. Having in mind the learning of songs as a sequence of notes (syllables), the problem of complex sequences arises immediately.

A basic observation is the presence of specific neurons, which are coding for the *transition* between syllables [23]. Sequence detecting neurons have also been identified in other systems, such as the visual system, or the auditory cortex of the bat [24]. These sequence detecting neurons can be used to produce sequences of order 1 : if a neuron T is coding for the transition from the note A to the note B, it can be used to provoque the transition from B to the note following (A,B). Sequences of higher order can be produced if we have also neurons which code for the transition between sequence detecting neurons. The architecture of the network is thus the following : one layer of neurons which code for the notes, and receive inputs from the external world ; a layer of "hidden" neurons, which code for the transitions. One starts with a high initial connectivity, and the learning procedure will select the usefull connexions. The precise dynamics and learning rules relies on Hebbian rules - that is the modification of a synaptic efficacy depends on the local activities of the neurons -, and more specifically on an interaction between three neurons. Suppose T has to detect $A \rightarrow B$. We want that the activity of B contributes to activate T only if A was active previously. This is achieved by the *synaptic triad* BTA : the synaptic efficacy J_{BTA} of the synapse $B \rightarrow T$ is potentiated by the activity of A. That is, if A is active J_{BTA} increases toward a maximal value, and decreases otherwise. Such heterosynaptic regulations [25] have been experimentally observed in several systems [26]. Learning will consist in modifying the maximal value of J_{BTA}, so that it can be large if the transition $A \rightarrow B$ does occur in the song, and negligeable otherwise.

To summarize, this model has made a first step towards a realistic learning procedure. The basic ideas however are the same, considering the way a sequence is produced : there are synapses responsible for the stabilization of the patterns, and synapses responsible for the transitions. An alternative to the sequence detecting neurons is to have sequence-detecting *patterns*, as shown by D. Amit [10].

Conclusion

I have not try to make an extensive review on the subject, but rather to present some basic ideas and models for processing temporal sequences in neural networks. In particular I tried to show where are the main difficulties which arise when one deals with sequences and not patterns : for asynchronous updating, the competition between stability of a pattern and transition to the next one has to be controlled ; whether it is for artificial or biological applications, the design of the network must be adapted to the degree of complexity of the sequences to be learned.

Acknowledgements

I had the pleasure to work on temporal sequences with J.P. Changeux, S. Dehaene, G. Toulouse, M. Opper, I. Guyon, L. Personnaz and G. Dreyfus. Discussions with M. Mézard and H. Gutfreund are gratefully acknowledged. I thank G. Weisbuch for a critical reading of the manuscript.

References

[1]   J.J. Hopfield, Proc. Natl. Acad. Sci. USA 79, 2554 (1982).

[2]   J.D. Cowan and D.M. Sharp, preprint (1987); D.W. Tank and J.J. Hopfield, Scientific American 257, 62 (1987), ; E. Bienenstock, F. Fogelman-Soulié and G. Weisbuch, "Disordered Systems and Biological Organizations", Springer-Verlag, Berlin 1986.

[3]   D.J. Amit, in Heidelberg Colloquium on Glassy Dynamics, J.L. van Hemmen and I. Morgenstern eds., Springer-Verlag, Berlin 1987, p. 430.

[4]   P. Peretto, J.J. Niez, in "Disordered Systems and Biological Organization", E. Bienenstock, F. Fogelman-Soulié and G. Weisbuch eds., Springer, Berlin 1986, pp 115-133.

[5]   H. Sompolinsky and I. Kanter, Phys. Rev. Lett. 57, 2861 (1986); D. Kleinfeld, Proc.Natl. Acad. Sci. USA 83, 9469 (1986).

[6]   S. Dehaene, J.P. Changeux, J.P. Nadal, Proc. Natl. Acad. Sci. USA 84, 2727 (1987).

[7]   L. Personnaz, I. Guyon, G. Dreyfus, Phys. Rev. A34, 4217 (1986).

[8]   G. Mitchison, talk given at Bad Homburg meeting on Brain Theory, 15th-19th September 1986.

[9]   J. Buhmann, K. Schulten, Europhys. Lett. 4, 1205 (1987).

[10]   D.J. Amit, Proc. Natl. Acad. Sci. USA (1988).

[11]   H. Gutfreund, M. Mézard, Phys. Rev. Lett., submitted.

[12]   I. Guyon, L. Personnaz, J.P. Nadal and G. Dreyfus, to appear in Phys. Rev A.

[13]   J. Keeler, 1986.

[14]   S. Venkatesh, Proceedings of the Conference on Neural Networks for Computing (Snowbird 1986); T.M. Cover, IEEE transactions EC14 3, 326 (1985).

[15] E. Gardner, Europhysics Lett. 4, 481 (1987).

[16] S. Diederich, M. Opper, Phys. Rev. Lett. 58, 949 (1987); G. Pöppel, U. Krey, Europhys. Lett. (1987); D. Kleinfeld and D.B. Pendergraft, Biophys. 51, 47 (1987).

[17] W. Krauth, M. Mézard, J. Phys. A20, L745 (1987).

[18] F. Rosenblatt, "Principles of Neurodynamics", Spartan Books, N.Y. 1962.

[19] T. Kohonen, "Self-Organization and Associative Memory", Springer-Verlag, Berlin (1984).

[20] L. Personnaz, I. Guyon, G. Dreyfus, Phys. Rev. A34, 4217 (1986).

[21] A. Albert, "Regression and the More-Penrose Pseudo-inverse" (Academic Press, New York, 1972)

[22] J.P. Nadal and M. Opper, unpublished.

[23] P. Marler, S. Peters, Dev. Psychobiol 15, 369 (1982); D. Margoliash, J. Neurosci. 3, 1039 (1983).

[24] A. Riehle and N. Franceschini, Exp. Brain Res. 54, 390 (1984); N. Nakayama, Vision Res. 25, 625 (1985); N. Suga in "Dynamic Aspects of Neocortical Function", G. Edelman, W.E. Gall and W.M. Cowan, Eds. (Wiley, New York, 1984) pp 315-373.

[25] T. Heidmann, J.P. Changeux, C.R. Acad. Sci. Ser. 2 295, 665 (1982).

[26] M. Ito, M. Sakurai and P. Tongroach, J. Physiol. (London) 324, 113 (1982); R. Hawkins, E. Kandel, Psychol. Rev. 91, 375 (1984).

HIERARCHICAL DIFFUSION[*]

CONSTANTIN P. BACHAS[**]

Stanford Linear Accelerator Center
Stanford University, Stanford, California 94305

ABSTRACT

We review the solution and properties of the diffusion equation in a hierarchical or ultrametric space [1].

One of the oldest equations in physics is the equation of diffusion. In its most general form it reads:

$$\frac{dP_i}{dt} = \sum_{j=1}^{N} \epsilon_{ij} P_j \tag{1}$$

where P_i is the probability of finding a particle at site i ($i = 1, ..., N$) of some arbitrary discretized space $\mathcal{M}$, and ϵ_{ij} is the hopping or transition probability per unit time, from site j to site i. This must of course be positive:

$$\epsilon_{ij} \geq 0 \qquad for \quad i \neq j \tag{2a}$$

and, in order to conserve total probability ($\sum_i P_i(t) = 1$), we must have :

$$\sum_{i=1}^{N} \epsilon_{ij} = 0 \tag{2b}$$

which fixes the diagonal elements ϵ_{ii}. Furthermore, I will assume in this talk that the transition matrix is symmetric:

$$\epsilon_{ij} = \epsilon_{ji} \tag{3}$$

i.e. that there is equal probability of hopping forward and backward between two sites. This restriction is not necessary for having a well defined diffusion problem, but as we will see it can be effectively lifted.

---

[*] Work supported by the Department of Energy, contract DE-AC03-76SF00515 .

[**] Permanent address: Centre de Physique Theorique, Ecole Polytechnique, 91128 Palaiseau, FRANCE.

Some general facts follow immediately from conditions (2-3). Firstly, there is a time-independent, steady-state solution, corresponding to equal probability at every site:

$$P_i(t) = \frac{1}{N} \qquad for\ all\ i \tag{4}$$

Secondly, for any vector $\mathbf{x}$: *

$$\mathbf{x}\epsilon\mathbf{x} = \sum_{i \geq j} \epsilon_{ij}(2\mathbf{x}_i\mathbf{x}_j - \mathbf{x}_i^2 - \mathbf{x}_j^2) \leq 0 \tag{5}$$

which shows that with the exception of the zero eigenvalue, corresponding to the vector (4), all other eigenvalues of the matrix ϵ are negative. Denote by

$$\epsilon^{(1)} = 0 > \epsilon^{(2)} \geq \ .... \geq \epsilon^{(N)} \tag{6}$$

these eigenvalues, and by $\mathbf{v}_i^{(1)} = \frac{1}{\sqrt{N}}$, $\mathbf{v}^{(2)},...,\mathbf{v}^{(N)}$ the corresponding orthonormalized eigenvectors. Knowledge of these clearly suffices to solve the diffusion problem (1), for any initial probability distribution $P(0)$ at $t = 0$. The result is:

$$P_i(t) = \frac{1}{N} + \sum_{I=2}^{N} < P(0)|\mathbf{v}^{(I)} > \mathbf{v}_i^{(I)} \ exp(-t/\tau^{(I)}) \tag{7}$$

where we have here defined the characteristic times :

$$\tau^{(I)} = -\frac{1}{\epsilon^{(I)}} \quad , \tag{8}$$

$< \mathbf{x}|\mathbf{y} >$ is the usual vector inner product, and we have used the fact that $< P(0)|\mathbf{v}^{(1)} >= \frac{1}{\sqrt{N}}$ since total probability is always one.

It follows easily from (7) that, at large t, one always approaches asymptotically the steady state (4). For finite spaces $\mathcal{M}$ the approach is exponential and dominated, generically, by the largest characteristic time $\tau^{(2)}$. For infinite spaces the relaxation can be slower than exponential, if there is a sufficient concentration of characteristic times at infinity. For instance if the density of characteristic times behaves asymptotically when $\tau \to \infty$ as:

$$\rho(\tau) \sim \tau^{-\nu-1} \tag{9}$$

then for generic initial conditions the relaxation is described at large times by a power law:
$\int^{\infty} \rho(\tau)e^{-\frac{t}{\tau}}d\tau \sim t^{-\nu}$.

---

* We will often drop the vector index labelling the sites of $\mathcal{M}$; summation over repeated indices is as usual implied.

Now we all learned in highschool how to solve the problem of diffusion on a *regular lattice* in D-dimensional Euclidean space. The eigenvectors of the transition matrix are in this case plane waves, their eigenvalue is proportional to momentum squared, and $\nu = \frac{1}{2}D$. Another situation in which one can, if not completely solve the problem, at least calculate exponents such as ν, is when the space $\mathcal{M}$ is scale-invariant or *fractal* [2]. Here I will consider yet another kind of geometry, *ultrametric* geometry, which has neither translational nor , necessarily, scale invariance [3] . Ultrametricity is the statement that , for all i, j and k :

$$\epsilon_{ij} \geq min(\epsilon_{ik}, \epsilon_{jk}) \tag{10}$$

which implies that given any three sites, the two smallest transition rates are equal. A more descriptive but completely equivalent way of saying this is that the sites can be organized as the leaves of some genealogical tree, so that transition rates are only a (decreasing) function of relation, i.e. a) $\epsilon_{ij} = \epsilon_{A(i,j)}$ is only a function of the nearest common ancestor $A(i,j)$ of i and j , and b) ϵ_A is monotone decreasing as A moves up, along any path, towards the patriarch or root of the tree. We may in fact encode all the information about the transition matrix ϵ in the tree, by stretching the heights h_A of its branch-points , so that $\epsilon_A \equiv e^{-h_A}$ (see fig. 1a). The height $h_{A(i,j)}$ can be thought of as an energy barrier, through which a particle going from site i to site j must penetrate. With this convention, both topology and branch heights will be relevant when we refer to trees in the sequel.

Much of the interest in ultrametric geometry was spurred by the discovery [4] that it describes the space of spin-glass states in mean field theory [5] . An exact or approximate hierarchical organization, however, also characterizes a wide variety of other natural and artificial systems. For the mathematicians and, more recently the string theorists, a familiar example of an ultrametric norm is the p-adic norm on rational numbers. What is surprising is that ultrametricity is powerful enough to allow us to solve the problem of diffusion exactly, without any further assumption or approximation [1] .

In order to describe the diagonalization of an arbitrary ultrametric transition matrix, let us introduce some notation (see also fig.1a): for any branch point or tree leaf B, we let B_n be its unique n-th ancestor, N_B the total number of its final descendants , i.e. tree leaves generated by B, and $\chi^{(B)}$ their characteristic function:

$$\chi_i^{(B)} = \begin{cases} 1, & \text{if i is a descendant of B} \\ 0, & \text{otherwise.} \end{cases}$$

Then for each B :

$$\mathbf{u}^{(B)} = \frac{1}{N_B}\chi^{(B)} - \frac{1}{N_{B_1}}\chi^{(B_1)} \tag{11}$$

is an eigenvector of the transition matrix, that describes the process of equilibration

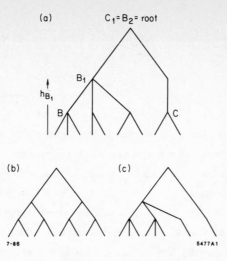

Figure 1 a) A generic tree illustrating our notation; the root is the father of B_1 and C, and the grandfather of B. The nodes B_1, B and C have 7, 3 and 2 final descendants respectively. All heights are measured from the leaves, which represent the sites of M. b) A self-similar, uniformly bifurcating tree. c) A most complex tree leading to slowest possible relaxation; its silhouette is the same as for (b), since total population doubles at every generation.

between the descendants of B and those of all his brothers. The corresponding eigenvalue, or inverse characteristic time, of this process can be expressed as a sum over all ancestors of B:

$$\epsilon^{(B)} = \sum_{n=1}^{root} N_{B_n}(\epsilon_{B_n} - \epsilon_{B_{n+1}}) \tag{12}$$

where by abuse of notation *root* stands here for the number of generations between B and the tree root, and all terms in eqs. (11) and (12) that refer to ancestors of the root should, by convention, be dropped. Suppose now that at $t = 0$, the particle is with probability one at a given site, i.e. tree leaf, L. We may decompose this initial condition in terms of the eigenvectors (11):

$$\delta_{i,L} = \sum_{n=0}^{root} \mathbf{u}^{(L_n)} \tag{13}$$

Since any other initial condition can obviously be written as a linear combination of (13), we have thus solved completely the problem of ultra-diffusion, *for any underlying tree.*

Let me now concentrate on the exponent ν that measures the speed of relaxation for infinite trees. Since there are many many more trees than real numbers, we expect some kind of universality. In other words ν should only depend on very few characteristics of the tree. One obviously relevant characteristic is the asymptotic rate of population growth, or *silhouette* :

$$s = \lim_{h,\Delta h \to \infty} -\frac{\Delta logn(h)}{\Delta h} \tag{14}$$

where $n(h)$ is the population at height h. A large s means that there are on the average more sites available for hopping in, at given barrier height, and this should speed up relaxation. Thus it makes sense to fix the silhouette s, and ask how ν depends on the remaining characteristics of the tree. For instance one may want to compare relaxation on the trees of fig. 1b and 1c, which have the same silhouette, i.e. would look the same if they were to wear a coat, but have completely different internal structure. The following three theorems address this issue. They hold under the assumption of stable relaxation, meaning that the autocorrelation function never decays faster than exponentially in time. I state these theorems without proof, since detailed proofs can be found in ref. [1]:

Theorem 1: *For uniformly multifurcating trees,* $\nu_{uniform} = \frac{s}{1-s}$. *For any other tree,* $\nu \leq \frac{s}{1-s}$.

Examples of uniformly multifurcating trees are the tree of fig.1b, or the tree of p-adic numbers. The above result shows that they lead to optimal relaxation. The next result identifies a class of trees leading to the slowest possible relaxation:

Theorem 2: *For the tree of fig. 1c,* $\nu = s$. *For any other tree,* $\nu \geq s$.

The third and last result shows that structural noise is irrelevant, in that it modifies the power law decay of autocorrelations by, at most, logarithmic corrections:

Theorem 3: *For trees whose branching ratio at every node is an identically distributed, independent random variable,* $\nu_{random} = \frac{s}{1-s}$.

Now both uniform and uniformly random trees are self-similar structures, whose parts are on the average identical to the whole. Thus we may interpret the above results as saying that , for fixed s, the dynamic exponent $-\nu$ measures the lack of self-similarity, or the *complexity* of the hierarchical structure. One can in fact define other, static measures of a tree's complexity, that share the same qualitative features as $-\nu$ [6,7] . An example is the critical threshold for percolation, p_c, from the root to the bottom of the genealogical tree. It can be shown [7] that

p_c is also minimized by self-similar trees, is insensitive to noise, and is maximized by the very unbalanced tree of fig. 1c. This suggests that ν may in fact be a monotone function of p_c, but I have not been able to prove this.

Let me make here a parenthetical comment on semantics. The use of the word *complexity* in the above context can be motivated in many ways. For instance it is in accordance with our intuitive notion that complex is neither ordered nor random [6]. And the threshold for percolation on the tree of winning strategies of a game is, indeed, a measure of the fault-tolerance and hence of the complexity of the game [7] . Nevertheless, I am fully aware of the fact that *complexity* figures high in the list of most used and abused words in the scientific literature (to give a very banal example: complex analysis is as we all know much simpler than real analysis). For this reason some other term for ν and p_c might have been more appropriate.

I will conclude with some remarks about where and how ultradiffusion could be usefull [8]. Clearly, since an ultrametric transition matrix allows for infinite range hoppings, we do not expect it to describe diffusive processes in a finite- dimensional space. $\mathcal{M}$ could however be either an infinitely connected artificial system , like a neural network, or the configuration space of a spin- or other statistical mechanical model . Consider in particular the mean-field spin glass [5]. It is reasonable to assume that like thermodynamic equilibrium states, long-lived metastable states also have a hierarchical organization [4]. Since the appearance of ultrametricity is, however, in this case spontaneous, we do not know a priori the structure of the underlying tree, which we need as an input in the diffusion equation. We could however try to work backwards; indeed, the hopping rates between metastable states are given by: $\epsilon_{ij} = e^{-\Delta F_{ij}/T}$, where ΔF_{ij} are free energy barriers. Let us make the naive assumption that the distribution of these barriers does not change, or changes very slowly with temperature. Then clearly the silhouette s is proportional to temperature, and the temperature-dependent dynamic exponent behaves like:

$$\nu(T) = \begin{cases} \frac{T}{T_c - T} & \text{for self-similar trees} \\ T & \text{for most complex trees} \end{cases} \tag{15}$$

below the critical temperature T_c, at which relaxation becomes unstable [1]. Note in particular that the transition to instability ($\nu = \infty$) is continuous in the case of self-similar trees, and discontinuous for the most complex ones. Now in the mean-field spin glass it is known [9] that below T_c:

$$\nu = \frac{1}{2} - \frac{T - T_c}{\pi T_c} + o(T - T_c)^2 \tag{16}$$

so that the transition is discontinuous. Thus, if we take the above naive model seriously, we would be tempted to conclude that the tree of the mean field spin glass is not self-similar. Interestingly enough, numerical efforts to reconstruct this tree give a result that looks very much like fig.1c [10].

Finally let me point out that the set of all ultrametric transition matrices sharing the same *topological tree*, is closed under both addition (trivially) and multiplication (we let the reader prove this for himself). Thus, both ultrametricity and the topology of the tree are stable under time-rescalings. It would be very interesting to study whether these stable trajectories in the space of all possible transition matrices, have any basins of attraction. In this case ultrametricity could be recovered as an effective property at large times. It would also be interesting to see whether the above ideas on complexity could be extended to the case of diffusion on multifractals [11].

REFERENCES

1. C.P.Bachas and B.A.Huberman , Phys.Rev.Lett. 57 (1986) 1965 ; J.Phys. A20 (1987) 4995 .

2. P.G. de Gennes, Recherche 7 (1976) 919 ; S.Alexander and R.Orbach, J.Physique Lett. 43 (1982) L625 ; R.Rammal and G.Toulouse, ibid 44 (1983) L13 .

3. Scale invariant ultradiffusion was introduced and studied before ref. [1], in different contexts and variations , by many authors : B.A.Huberman and M.Kerszberg, J.Phys. A18 (1985) L331 ; S.Teitel and E.Domany, Phys.Rev.Lett. 55 (1985) 2176 and 56 (1985) 1755 ; A.Maritan and A.L.Stella, ibid 56 (1986) 1754 and J.Phys. A19 (1986) L269; S.Grossman, F.Wegner and K.H.Hoffmann, J.Physique Lett. 46 (1985) L575 ; G.Paladin, M.Mezard and C.De Dominicis, ibid 46 (1985) L985 ; M.Schreckenberg, Z.Phys. B60 (1985) 483 ; A.T.Ogielski and D.L.Stein, Phys.Rev.Lett. 55 (1985) 1634 ; D.Kumar and S.R.Shenoy, Solid State Comm. 57 (1986) 927 ; A.Blumen, J.Klafter and G.Zumofen, J.Phys. A19 (1986) L77 .

4. M.Mezard, G.Parisi, N.Sourlas, G.Toulouse and M.Virasoro, Phys.Rev.Lett. 52 (1984) 1156 and J.Physique 45 (1984) 843 ; for a review of ultrametricity see also R.Rammal, G.Toulouse and M.A.Virasoro, Rev.Mod.Phys. 58 (1986) 765 and references therein .

5. D.Sherrington and S.Kirkpatrick, Phys.Rev.Lett. 35 (1975) 1792 .

6. B.A.Huberman and T.Hogg, Physica 22D (1986) 376 ; B.A.Huberman and H.A.Ceccatto, Xerox PARC preprint (1987) .

7. C.P.Bachas and W.F.Wolff, J.Phys.A20 (1987) L39 .

8. For a discussion of this issue see : R.G.Palmer, in Proceedings of the Heidelberg Colloquium on Glassy Dynamics and Optimization, eds. J.L.van Hemmen and I.Morgenstern, Springer Verlag 1986 and references therein, and also ref.[3].

9. H.Sompolinsky and A.Zippelius, Phys.Rev.Lett 47 (1981) 359 and Phys.Rev. B25 (1982) 6860 .

10. N.Parga, CERN-TH 4410 preprint (1986) .

11. G.Paladin and A.Vulpiani, Anomalous Scaling Laws in Multifractal Objects, Physics Reports, to appear.

RANDOM WALKS, RANDOM SURFACES, AND COMPLEXITY[1]

Ph. de Forcrand[a], F. Koukiou[b] and D. Petritis[b]

a. Cray Research, 1333 Northland Dr., Mendota Heights MN 55120, USA
and Physics Dept., University of Minnesota, Minneapolis MN 55455, USA
b. Institut de Physique théorique, Université de Lausanne, CH-1015 Lausanne

1 Introduction

The notion of complexity has not yet recieved a precise, generally accepted meaning, therefore it cannot be quantified properly. However, when comparing two different systems, very often one can say which is the more "complex", thus giving to the word complex an intuitive meaning. In fact, as stressed out by different speakers in this Meeting, there are many kinds of complexity : algorithmic, phase space, combinatorial etc. Since there is not yet a "Theory of Complexity" it is instructive to perform a case study of systems, relatively well understood, having different intuitive complexities and stress out which are the common trends between them and which are the special features that give to such systems different degrees of complexity.

Here, as examples of systems with different intuitive complexities we consider random walks and random surfaces. The advantage of looking at such systems is that many things are known analytically and their complexity is well controlled. For instance we know that counting these objects is a NP-complete problem but their phase space does not have this more than astronomical vastness typical of "conventional" complex systems. Hence they can be studied by numerical simulation.

Moreover, random walks (RW) and random surfaces (RS) are interesting to study in connection with physical models and not only as case studies of complexity. In this talk, some definitions are first given of what is meant by RW and RS and the following questions are addressed:

i) Why are RW and RS interesting?

ii) What are the theoretical problems one is faced with?

iii) Which methods can be used and what are the specific results obtained with each method?

iv) What are the open problems that remain to be solved?

---

[1]Work partially supported by the Swiss National Science Foundation. This text provides a slightly modified version of the talk given at the Meeting.

Some particular emphasis will be given on the methods we used to study RW[10,11,12] and (in collaboration with J. Ambjørn) RS [3].

The talk is organized in two independent parts presented in order of increasing "complexity", namely first RW and then RS are treated.

At the end we present some thoughts about relative complexity.

2 Random Walks

2.1 Definitions

We consider mainly walks on a d-dimensional lattice with coordination number q, but q is not necessarily restricted to be $q = 2d$ (cubic lattice). Other kinds of lattices are allowed, like hexagonal, trigonal, diamond etc.

A random walk on a lattice is the trajectory of a stochastic process, indexed by an integer time, that assigns a probability to the appending of a lattice bond at the end of the previous walk, while respecting some constraints.

For the ordinary random walk (ORW) no constraints have to be respected, hence every lattice site can be visited an arbitrary number of times by the walk. This lack of constraints gives to the ORW its Markovian character and makes its study considerably easier than the study of the other kinds of random walks introduced below.

A self avoiding random walk (SAW) respects the constraint that it can visit each lattice site at most once. It is also called 1-tolerant walk as a special case of the k-tolerant walk that can visit each site at most k times.

For the Edwards walk there is no constraint but only a damping in the case where it has many self-intersections i.e. a weight is attached on each such walk ω of the form $\exp(-\lambda I(\omega))$ where $I(\omega)$ is the number of self-intersections of the walk. One expects intuitively that "$\lim_{\lambda \to 0}$" Edwards = ORW and "$\lim_{\lambda \to +\infty}$" Edwards = SAW and it is a remarkable fact that these limits can be given a rigorous meaning.

The reader must be aware that the names used in this talk are not the standard ones used by the community of polymer physicists. We stick to this more economic nomenclature; the definitions given above make clear what is meant by ORW, SAW and Edwards walk. Another point to be stressed is that these random walks are objects existing independently of the methods used to simulate them. In particular, ORW and SAW depend uniquely on the lattice coordination number q and the dimensionality d.

2.2 Why random walks are interesting

The first walks historically studied were the ordinary random walks. They provide a discretized version of the Brownian motion, hence they can be used as a laboratory to check the validity of many ideas. They can also be used as rough models for polymers. A great revival of the interest in ORW arose when Symanzik introduced a representation of scalar quantum field theory in terms of random walks [25].

The SAW's were studied for a long time by chemists as quite realistic models of polymers [14]. Their interest for field theory was pointed out by de Gennes [19] when he realized that they arise in the representation of vector quantum field theory with interaction term $(\vec{\phi} \cdot \vec{\phi})^2$ in the limit where the number of vector components tends to zero. Although this limit sounds quite strange and non physical, it proves very rich in structure and can be given a precise meaning by analytic continuation in the number of components.

In many respects the Edwards walk is very interesting. First from a theoretical point of view it provides a "continuous interpolation" between ORW and SAW. The study of the limits is a challenging theoretical problem. Moreover, λ being a free parameter, choosing it negative corresponds to self-attracting polymers—a case that occurs if the polymer develops attracting forces between its parts e.g. van der Waals or hydrogen bond like attractions.

Mathematicians have introduced related objects like Wiener sausages (i.e. locus of points described by a ball whose center runs over an ORW), excluded volume SAW's (i.e. SAW's with the additional constraint that the sausage obtained by letting a rigid impenetrable ball run on the SAW can occur), etc. These mathematical constructions provide even more realistic polymer models than random walks.

To summarize, some random walks belong to universality classes shared by many physical systems (i.e. field theory, statistical mechanics or chemical physics); it is however easier to study random walks than the underlying physical model!

2.3 Theoretical Problems

The study of RW on a lattice is fundamentally an enumeration problem. As for most enumeration problems on graphs, counting specific subclasses of RW on a lattice is very often a NP-complete problem. Useful quantities to study are:

- Number of walks c_N, starting at the origin, ending anywhere on the lattice and having length N.

- Number of walks $c_{N,x}$, starting at the origin, ending at point x of the lattice and having length N.

- Mean gyration radius $\langle r \rangle_N$ defined as the mean end-to-end Euclidean distance of walks having length N.

- Number of pairs $d_{N,N}$ of independent inresecting walks starting any two sites, both having length N.

These enumeration problems have distinct features for the different kinds of walks. For the ORW, $c_N = q^N$ and $c_{N,x}$ and $\langle r \rangle_N$ can be computed analytically using combinatorial arguments [21]. For the SAW, the determination of these very quantities is a NP-complete problem! For the Edwards walk, one needs also to know the number of self intersections. The determination of the quantity $d_{N,N}$ is an unsolved problem even for the simplest case of ORW.

Although the exact enumeration presents such a contrasted behaviour for different kinds of RW, it is a remarkable fact that the asymptotic behaviour of these quantities for large N has an universal form i.e.

$$c_N \underset{N \to \infty}{\sim} \mu^N N^{\gamma-1}$$

and

$$c_{N,x} \underset{N \to \infty}{\sim} \mu^N N^{\alpha-2}.$$

The exponential behaviour μ^N is governed by the effective coordination number μ and the power law behaviour by a critical exponent γ or α. One expects that $\mu(\text{SAW}) \leq \mu(\text{Edwards with } \lambda > 0) \leq \mu(\text{ORW}) = q$. Numerical simulation confirms that $\mu(\text{SAW}) < q$.

The mean gyration radius $\langle r \rangle_N$ plays an important role in determining the geometry of the trajectories of RW; its asymptotic behaviour

$$\langle r \rangle_N \underset{N \to \infty}{\sim} N^\nu$$

is governed by the critical exponent ν. It is generally accepted that the fractal dimension d_F of the trajectory is $d_F = 1/\nu$. For the ORW, the critical exponent ν is proven to be equal to $1/2$ in any dimension d. For the SAW, this exponent varies with d and equals to $\nu = 1$ in $d = 1$, $\nu = 3/4$ in $d = 2$, $\nu = 0.5745$ in $d = 3$, and tends asymptotically for large d to $\nu = 1/2$. For the Edwards walk, this critical exponent interpolates between the corresponding values for ORW and SAW. This problem is actually under investigation by our group using Monte Carlo simulation [13].

All the previous quantities refer to individual properties of walks. The last one, namely the number $d_{N,N}$, refers to pairs of walks and is in many respects much more difficult to determine. This number has a direct relationship with the renormalized coupling constant in quantum field theory. Its asymptotic behaviour

$$d_{N,N} \underset{N \to \infty}{\sim} \mu^{2N} N^{2\Delta_4-\gamma}$$

exhibits an exponential part μ^{2N} and a power law $N^{2\Delta_4-\gamma}$ associated with the linear combination of critical exponents Δ_4 and γ.

All these critical exponents $\gamma, \nu, \alpha, \Delta_4$ are not linearly independent, actually there are two relations linking them together known as hyperscaling relations. They read

$$d\nu = 2 - \alpha$$
$$d\nu = 2\Delta_4 - \gamma.$$

The second hyperscaling relation has a very controversial history. Using a high-temperature series expansion, Baker [4] claimed that hyperscaling fails for the Ising model in more than three dimensions. Using an argument similar to that used by Aizenman [1] to prove hyperscaling for the two dimensional Ising model, one can only prove that $d\nu - 2\Delta_4 + \gamma \leq 0$ for the two dimensional SAW. Des Cloizeaux argued that hyperscaling must fail if $d\nu > 2$ and Sokal proved it [24]. However, this last inequality is believed to occur in $d \geq 4$ only. Much of our subsequent work was motivated by trying to confirm this hyperscaling relation for the SAW in three dimensions. The rigorous proof of hypercaling for the SAW in three dimensions still remains an open problem.

2.4   Methods of study and results

Various methods are used to study RW; they are complementary.

i) The first method used historically is probability theory. For ORW there is a huge number of results. A recent extensive bibliography can be found in [18]. For the Edwards walk in $d = 2$, Varadhan [26] proved that the Edwards walk follows exactly the same trajectories as the ORW! This very remarkable result is in fact proven for the continuum version of the renormalized Edwards walk where it is shown that the measure of Edwards walk has a finite density w.r.t. the Wiener measure. It took 11 years to realize that this result does not extend to $d = 3$. In fact Westwater [27] proved that the Edwards walk does not follow the same trajectories as ORW in 3 dimensions. Some results are also known for Edwards in higher dimensions $d \geq 5$ [8] and asymptotically for $d \to \infty$ [23]. For the SAW, the use of probability theory is very difficult since they don't arise as Markov processes. Some results concerning loop-erased ORW are only known [15,16,17].

ii) A second method that can be used is exact enumeration followed by extrapolation. For ORW of course this is a more or less trivial problem since analytic formulae exist for c_N and $c_{N,x}$. For SAW, some exact enumerations have been done up to walks of length $N \sim 20$ and then the extrapolation to $N \to \infty$ is studied using Padé approximants. Although this method gives more or less reliable estimates for μ, it is quite biased for the critical exponents. Finally, for the Edwards walk this method does not apply as such; one needs also to count the number of self intersections and the problem rapidly becomes non tractable.

iii) One can use the analogy of RW with quantum field theory that was stressed in the previous section, apply the techniques of the renormalization group to the underlying field theory problem and extract the critical exponents. This method cannot be used to test the hyperscaling relations since they are implicitly assumed valid in the renormalization group approach.

iv) The simulation methods remain as a last resort; they are used mainly for SAW's and Edwards walks. Such simulations are feasible because the phase space of all possible walks has quite a reasonable size, namely it does not grow more than exponentially with the length of allowed walks as reflected in the asymptotic behaviour of c_N. There are various simulation procedures.

a) Try and reject algorithms: Simulate ORW and reject all non-SAW in the sample. This simulation procedure cannot generate long walks. Since we know that $c_N(\text{SAW}) \sim \mu^N N^{\gamma-1}$ and $c_N(\text{ORW}) = q^N$ only a $(\mu/q)^N$ part of SAW of length N survive in the sample. For all practical purposes, N cannot be taken bigger than 25.

b) Dynamical algorithms: We distinguish between growth algorithms where non equilibrium statistical mechanics are needed to analyze the results (see [20] for a recent review) and equilibrium algorithms where only equilibrium statistical physics are needed since we have the guarantee of attaining a unique equilibrium probability distribution.

It is worth noting that there is no universal simulation procedure i.e. every simulation is optimized for obtaining a particular subclass of exponents. For a given procedure, it may be either impossible to obtain the remaining exponents or we can obtain them with conditions far from optimality.

2.5   The particular algorithm

In the following we describe in some detail the particular algorithm we used in our simulations. This is a very simple dynamical algorithm for generating RW. A closely related algorithm was introduced in [22] and in its present form in [5]. It goes as follows:

Begin from the empty walk anchored at the origin.
Choose a parameter β (monomer activity).
Repeat many times:
{ Choose a random number uniformly distributed in [0,1].
If $r > (1 + q\beta)^{-1}$
   then try to append a link in one of the q directions at the end of the walk;
   If the resulting walk is SAW
      then effectively append the link;
      else consider the previous walk once more.
   else try to delete the last bond;
   If the walk is not empty
      then effectively delete the last bond;
      else consider the empty walk once more.}

This algorithm has very simple elementary moves, is Markovian in the large space of all possible SAW's, fulfils the detailed balance condition, and is ergodic. The last two conditions guarantee that a unique equilibrium probability distribution is attained. The statistical properties of this algorithm are well understood. In fact, the generated walks are correlated but the autocorrelation time τ behaves in a simple manner $\tau \sim \langle N \rangle^2$. Intuitively, this is understood since an arbitrarily long walk can decay to the empty walk with finite probability in approximately $\langle N \rangle^2$ steps. Once the empty walk is attained all previous memory is lost.

In order to take advantage of the low autocorrelation time $\tau \sim \langle N \rangle^2$, the time needed for the test of the self-avoiding condition must not depend on the actual length of the walk. This fact dictates as data structure a bit-map table of site occupation that is updated in every elementary move.

A particular attention is paid to the fit and error analysis. Without reproducing the details that can be found in [5,10,12] let us mention that we used the maximum likelihood fit. The naive statistical errors are multiplied by $\sqrt{\tau}$ to get rid of correlation in the sample. Systematic errors arising as finite size corrections to the asymptotic laws are taken into account phenomenologically. The algorithm used is optimal for the determination of μ and of the critical exponents ν and γ. It is rather poorly adapted for the determination of Δ_4 and α. The values obtained for $\mu, \gamma, \nu, \alpha, \Delta_4$ are quoted in the original papers in $d = 2$ and 3. Here we only give the value of $d\nu - 2\Delta_4 + \gamma = -0.0082 \pm 0.027 \pm 0.018$ (where the error bars represent the statistical and systematic errors) in $d = 3$. The point 0.0 lies within less than a standard deviation from the central value, hence the hyperscaling relation is supported in 3 dimensions. This remark closes the long controversy about the failure of the hyperscaling in 3 dimensions.

2.6 Open problems

In the light of the results obtained by Varadhan [26] and Westwater [27] it is not yet clear whether the generally accepted relationship between fractal (Hausdorff) dimension d_F and the inverse of the critical exponent ν always holds. In particular, there is a lack of understanding for the Edwards walk. Both simulations and mathematical investigations are in progress to clarify this topic.

Another problem that remains to be understood is the statistical physics of interacting random walks—a special case of statistical physics of extended objects. But an extended one dimensional object evolving in time describes a random surface. This remark provides a smooth transition to the second part of this talk.

3 Random surfaces

3.1 Definitions

Random surfaces can be viewed either as fluctuating two-dimensional objects (e.g. interface between two different media) or as the world sheet of a topologically one-dimensional object (e.g. a string or a RW evolving in time). A random surface is a manifold having many different characteristics:

i) Global characteristics such as: the number of handles g taking any non negative integer value; the existence or absence of boundary; the orientability etc.

ii) Local characteristics like: local curvature and local twist both being described by the metric tensor field $g_{\mu\nu}(\xi)$.

iii) Embedding characteristics i.e. we assume that on the surface lives a field $X(\xi)$ taking its values in an embedding space E. We distinguish various possibilities describing different physics e.g. $E = \{e\}$ provides a trivial embedding, $E = \{-1,1\}$ gives a spin system on the surface, $E = \mathbb{R}^d$ describes a string, $E = \mathbb{R}^d \otimes \{-1,1\}^f$ a superstring etc.

iv) Discretization characteristics: a RS has at least two different *nonequivalent* discretizations. They are obtained either by gluing plaquettes of $\mathbb{Z}^3$, or by simplicial decomposition (triangulation).

In the following, we shall consider mainly surfaces with the topology of the sphere, embedded in $\mathbb{R}^d$, and discretized by simplicial decomposition.

3.2   Interest of RS

Random surfaces prove very useful in many different branches of pure and applied science. In solid state physics they arise as interfaces between two different media e.g. droplets, solid-solid, solid-liquid interfaces etc. In elementary particle physics, RS are connected with many aspects of string theory, namely as bosonic strings (see below), fermionic strings or even as superstrings (which are expected to be the ultimate theory of nature). In engineering they arise as models for vibrating membranes, in computer science in relation with finite elements methods, in optimization theory for some special properties of graphs, in probability theory as generalizations of random processes on manifolds etc.

3.3   Theoretical problems

The theoretical problems one is called to solve when dealing with RS are in many respects similar to the problems encountered with RW, namely counting problems. As in the RW, exact counting of RS is a NP-complete problem but the asymptotic behaviour is quite well understood. For instance, the number of surfaces c_N, with spherical topology, having N triangles behaves as $c_N \underset{N\to\infty}{\sim} \mu^N N^{\gamma-2}$ and the mean gyration radius $\langle r \rangle_N$ behaves as $\langle r \rangle_N \underset{N\to\infty}{\sim} N^\nu$, the fractal dimension, d_F, of the surface being given by $d_F = 1/\nu$. A remarkable property of the critical exponents γ and ν is their dependence on the dimensionality, d, of the embedding space $\mathbb{R}^d$.

Since RS are topologically 2-dimensional objects, they present some new features, absent in RW. These features are gathered under the category of optimization problems. For instance one can think of finding the geodesic path between two arbitrary vertices of the surface or of counting the number of spanning trees. This last problem, although generally NP-complete in the number, N, of vertices, has a particularly simple (in fact N^3) exact analytical solution. For a given N the number of spanning trees t_N on a triangulated spherical surface is *equal* to the determinant of the incidence matrix truncated by one row and one column.

3.4 Methods used and specific results

One can use the same methods to study RS as for RW. However some of them give very poor results. The probability theory for stochastic processes on manifolds is in a embryonic age, the exact enumeration/extrapolation method is practically inapplicable since one can enumerate surfaces up to 8 triangles only, the field theory methods are of little help since the underlying field theory is a string theory. The last resource that remains is Monte Carlo simulation. Now one is faced with the problem of inequivalent discretizations. If we discretize by gluing plaquettes of $\mathbb{Z}^3$ we get a trivial limit because this kind of surfaces degenerates to branched polymers i.e. for entropic reasons the surfaces that dominate in the sample are thin tubes with volume approximately equal to their area [7].

The only remaining (seemingly nontrivial) discretization is the simplicial decomposition. Triangulated surfaces are simulated in canonical [6] or grand canonical ensembles [3,9] and the various critical exponents ν, γ are obtained as functions of the embedding space dimensionality.

3.5 The particular algorithm

The choice for this particular algorithm was dictated by our wish to simulate RS arising as discretized versions of the Polyakov string with partition function (in the continuum) given by

$$Z(\beta) = \int Dg_{\alpha\beta} \int D\vec{x} \, \exp(-\beta \int d^2\xi \sqrt{g} \partial_\alpha x^\mu(\xi) \partial^\alpha x_\mu(\xi))$$

with $\alpha, \beta = 1, 2$ and $\mu = 0, 1 \ldots, d-1$.

As discretized version we take

$$Z(\beta) = \sum_{T \in \mathcal{T}} \frac{1}{S_T} \rho(T) \int (\prod_{i \in T} d^d x_i) \, \delta(\sum_{i \in T} \frac{x_i}{|T|}) \exp(-\beta \sum_{\langle i,j \rangle} (x_i - x_j)^2).$$

The transcription from the continuous formula to the discrete one is quite obvious: the integration over all possible metrics is substituted by a summation over all non-singular triangulations $\mathcal{T}$ and $\partial_\alpha x^\mu \partial^\alpha x_\mu$ by the discrete Laplacian on the surface. Finally, $\rho(T)$ is a factor coupling to the local curvature and S_T a symmetry factor. The gaussian integration over x is performed explicitly giving

$$Z(\beta) = \sum_{T \in \mathcal{T}} \frac{1}{S_T} \rho(T) \, (\det I'_T)^{-d/2}$$

where I'_T is the truncated incidence matrix [2]. It is this last form of partition function that determines the probability distribution in the sample generated by our algorithm. Now, the details of the algorithm are easy to catch:

Begin from the tetrahedron.

Repeat many times:

{ Choose a random number r uniformly distributed in $[0,1]$.

If $r > 1/2$

    then try to insert a vertex v' i.e. choose a vertex v on the triangulation and two of its neighbors i_1 and i_2 at random. This local configuration defines the situation A. Now, delete the bonds $(v, i_1+1), \ldots, (v, i_2-1)$; insert the vertex v' and add the bonds $(v', v), (v', i_1), \ldots, (v', i_2)$. This local configuration defines situation B. Compute the corresponding truncated incidence matrix I'_A and I'_B and the quantity $q = (\det I'_A / \det I'_B)^{d/2} \times$ symmetry factors.
Choose $q' \in [0,1]$ at random.
If $q' > q$

        then accept situation B;

        else keep situation A once more.

    else try to delete a vertex i.e. choose a vertex v and one of its neighbors m at random. This local configuration corresponds to situation A. Delete all the bonds connecting v to its neighbors and connect m to all neighbors of v that are not neighbors of m. This local configuration corresponds to situation B. Compute the truncated incidence matrices I'_A and I'_B and the quantity $q = (\det I'_A / \det I'_B)^{d/2} \times$ symmetry factors.
Choose $q' \in [0,1]$ at random.
If $q' > q$

        then accept situation B;

        else keep configuration A once more.}

Remark: The "insert step" of the previous algorithm is always implementable. For the "delete step" however one must check in addition that i) one does not create a surface with fewer than 4 triangles and ii) no loops of length 2 are created. These two conditions are easily implemented by an additional check in the delete part, omitted here for clarity.

This algorithm fulfils the detailed balance condition and is ergodic, therefore it generates a unique equilibrium probability distribution. The statistical properties of the algorithm are well understood: subsequent surfaces are correlated but all memory is lost once we attain the tetrahedron. The data structure used to keep track of surfaces is obvious: an incidence matrix, I, updated in every step and a system of pointers giving the ordinal numbers of neighbors of each vertex. We did not try to optimize this data structure because for large N, 97% of the simulation time is consumed in the computation of the determinant.

The critical exponent γ is extracted by a maximum likelihood fit and the naive statistical errors are amplified by $\sqrt{\tau}$ (τ = autocorrelation time). The systematic errors are taken into account phenomenologically.

Figure 1 summarizes the results obtained for γ as a function of the embedding space dimension d. The error bars are total (i.e. statistical and systematic) errors. The simu-

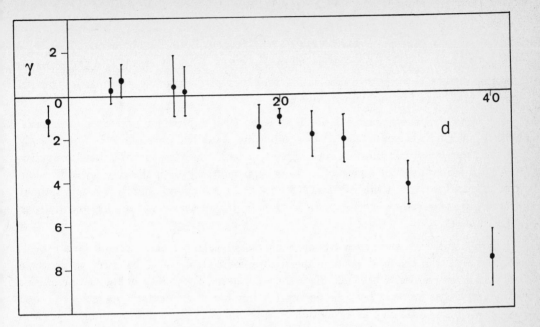

Figure 1: The values of the critical exponent γ as a function of the dimension d of the embedding space.

lations needed to produce figure 1 were performed on different Cray computers and the total XMP-equivalent CPU time is of the order of 500 hours. Hence, it seems unfeasible to decrease substantially the error bars using computers of the present generation.

3.6   Open problems

The main question that in our opinion remains open in RS is the possible existence of an interval $[d_\ell, d_u]$ of embedding space dimensions where $\gamma(d) > 0$. Positivity of $\gamma(d)$ for dimensions, d, belonging to the interval delimited by the lower, d_ℓ, and the upper, d_u, critical dimensions means that a continuum limit can be defined for these dimensions; hence, the triangulated RS provide a sensible discretization of the world sheet of the bosonic string. However, Figure 1 does not allow to assert that $\gamma(d) > 0$ for any dimensions d.

We have already seen that discretization by gluing plaquettes degenerates to branched polymers yielding a trivial limit. If $\gamma(d) < 0$ for every d, it means that even the simplicial decomposition does not provide a sensible discretization, hence some new ideas are needed. If it turns out that this new discretization does not degenerate, it will give *the* discretized version of the string.

4 Conclusions

Exact enumeration of both RW and RS is a NP-complete problem, a feature shared by many combinatorial or optimization problems (e.g. traveling salesman). Thus an exact solution of the problem is hopeless!

If instead of exact solutions one is satisfied with approximate or asymptotic ones, statistical methods can be used. But equilibrium statistical physics can be applied only if the thermodynamic limit exists. This happens if the partition function (or equivalently the phase space) increases at worst exponentially with the degrees of freedom (or the volume). Both the RW and RS have a thermodynamic limit. We speak about *statistical complexity* when the partition function diverges worse than exponentially in the volume.

Now, statistical physics methods being possible, one must use appropriate algorithms with the required good properties (ergodicity, detailed balance) to generate statistical ensembles. Once such algorithms are derived, one may define two quantities M_N and S_N as follows: Let M_N denote the number of elementary operations needed for one Monte Carlo step as a function of the number of degrees of freedom N, asymptotically for large N. e.g. $M_N \sim 1$ for SAW and $M_N \sim N^3$ for RS. Let S_N denote the number of MC steps needed to produce two statistically independent events with the given algorithm. e.g. $S_N \sim N^2$ for both SAW and RS. The *computational complexity*— to be distinguished from the algorithmic complexity—is the asymptotic behaviour for large N of the product of these two quantities. What makes RS to look more "complex" than RW is that this number goes as $N^3 \times N^2 = N^5$ for the algorithm used for RS and $1 \times N^2 = N^2$ for RW. This gives the relative complexities of the problems when treated with the given algorithms. However, this provides a comparison of algorithms and not of problems.

The *algorithmic complexity* of a problem must be defined in a more intrinsic way; as a tentative definition one can take the infimum over all algorithms of the computational complexity. What a case study provides is an upper bound of the algorithmic complexity.

In conclusion what this case study teaches us is that

i) we must distinguish between different complexities, let us call them mathematical, statistical and algorithmic in our case.

ii) Algorithmic complexity (in the sense introduced above) has a meaning only if the statistical complexity is not worse than exponential.

iii) If the statistical complexity is worse than exponential we can say nothing in the light of the examples studied here, and some additional case studies must be performed.

References

[1] M. Aizenman, Commun. Math. Phys. **86**, 1 (1982)

[2] J. Ambjørn, B. Durhuus, J. Fröhlich, P. Orland, Nucl. Phys. **B270**[FS16], 457 (1986)

[3] J. Ambjørn, Ph. de Forcrand, F. Koukiou, D. Petritis, Phys. Letts. **B197**, 548 (1987)

[4] , G. Baker, Jr., Phys. Rev. **B15**, 1552 (1975)

[5] A. Berretti, A. D. Sokal, J. Stat. Phys. **40**, 483 (1985)

[6] A. Billoire, F. David, Nucl. Phys. **B275**[FS17], 617 (1986)

[7] A. Bovier, J. Fröhlich, U. Glaus, Mathematical aspects of the physics of disordered systems Lecture 5 of the Course given by J. Fröhlich at Les Houches Summer School, in *Critical Phenomena, Random Systems and Gauge theories*, K. Osterwalder and R Stora, eds. North- Holland, Amsterdam (1986)

[8] D. Brydges, T. Spencer, Commun. Math. Phys. **97**, 125 (1985)

[9] F. David, J. Jurkiewicz, A. Krzywicki, B. Petersson, *Critical exponents in a model of dynamically triangulated random surfaces*, Preprint LPTHE 87/10

[10] Ph. de Forcrand, F. Koukiou, D. Petritis, J. Stat. Phys. **45**, 459 (1986)

[11] Ph. de Forcrand, F. Koukiou, D. Petritis, Phys. Letts. **B189**, 341 (1987)

[12] Ph. de Forcrand, F. Koukiou, D. Petritis, J.Stat. Phys. **49**, 223 (1987)

[13] Ph. de Forcrand, F. Koukiou, D. Petritis, *Study of the Edwards random walk using Monte Carlo simulation*, in preparation.

[14] J. Mazur, Non-self-intersecting random walks in lattices with nearest neighbors interactions, in *Stochastic processes in chemical Physics*-Vol. XV, K. Shuler ed., Interscience, New York (1969)

[15] G. F. Lawler, Duke Math. J. **47**,655 (1980)

[16] G. F. Lawler, Commun. Math. Phys. **86**, 539 (1982)

[17] G. F. Lawler, Commun. Math. Phys. **97**, 583 (1985)

[18] J.-F. Le Gall, Thèse d'État, Paris (1987)

[19] P. G. de Gennes, Phys. Letts. **38A**, 339 (1972)

[20] L. Peliti, Riv. Nuovo Cim. **10**(6), 1 (1987)

[21] J. K. Percus, Combinatorial Methods, in *Applied Mathematical Science*, Vol. 4, Springer, New York (1971)

[22] S. Redner, P. J. Reynolds, J. Phys. A: Math. Gen., **14**, l5 (1981)

[23] G. Slade, Commun. Math. Phys. **110**, 661 (1987)

[24] A. D. Sokal, unpublished result.

[25] K. Symanzik, Euclidean Field Theory, in *Proceedings International School of Physics "Enrico Fermi"*, Varenna Course XLV, R. Jost ed. (Academic Press, New York, 1969)

[26] S. R. S. Varadhan, Appendix to the course given by S. Symanzik, *op. cit.*

[27] M. J. Westwater, Commun. Math. Phys. **72**, 131 (1980)

COMPLEXITY IN LARGE TECHNOLOGICAL SYSTEMS

Giovan B. Scuricini
ENEA-Dipartimento Reattori Veloci
CRE Casaccia C.P. N.2400 -00100 Roma A.D.-Italia

1. INTRODUCTION

A lot of attention has been paid in these very last few years (1, 2, 3, 4) to complexity in very large systems - economics, management, social, political - and many evaluation procedure have been proposed. The dimensions of these systems and their same nature limit the possibility of a quantitative or a precise qualitative evaluation ; these systems are thought to be practically unforeseeable because of subjective or objective (butterfly effect) indetermination principles. With a few exceptions (5,6) no quantitative measure - suitable for system engineering - has been proposed.

At the same time in the field of AI many theories have been elaborated to evaluate computational or algebraic complexity; the basic assumptions - convexity and linearity - limit the domain of application to the real world of engineering and the language adopted is far from the ones to which the engineer is used to.

Today is a common opinion among system engineers that complexity is one of the main cause of the present difficulties in plant design and control : the procedures established on newtonian paradigms - perfectly adequate for the old simple plants - are completely inadequate for the modern complex plants. The search for a measuring tool is not anymore a speculative desire but an urgent necessity; the search for simplicity can not be performed if at least a comparative evaluation of complexity cannot be made.

It is quite funny that nearly 20 years after the H. Simon famous example of the two watchmakers - Cronos and Tempus - so little attention is paid to an engineering science of complexity applied to "large scale plants" (7) , Casti theories (8) are not widely applied and he himself points out (p.106) that "computational complexity is much too restrictive for system-theoretic work" and that "the measurement of complexity for a system governed by the *non linear differential equation dx/dt= f(x,u)* is a considerable more difficult problem" (p.119). The main point may be that Casti, like the same

Simon , are giving a particular emphasis to hyerarchical structures and that real systems are non linear especially for the aspects that today concern system engineers. In the example quoted by Gottinger (p.126) the structural complexity may be of the order of 240^{10}; the modularization of the system may reduce the central control complexity but may introduce brand new problem of communication : this aspect is today , as we shall see, very important for Large Scale Plants.

We may say that reliability, safety, control, quality production depend on the solution given to the different problems raised by complexity (9, 10, 11, 12, 13, 14, 15).

We may remind that E.R. Caianiello says that : the laws that determines "structures" in physics are <u>not</u> of purely <u>physical</u> nature , such as Newton's and Maxwell's laws : they are rather <u>logical</u>, connected with information, entropy, the structuring of complex systems into levels" (16).

Before to introduce a proposal to evaluate the Large Technological Plants complexity I shall elucidate the reasons why we need today such an evaluation - which was not so much required in the past- and which are the different aspects of complexity that have to be taken in account. I shall try to shaw that it could be seriously misleading to evaluate complexity by a single parameter (or number); the simplification that should be introduced to do so could completely falsify the model .

2. THE COMPLEXITY INCREASE IN NATURAL AND ARTIFICIAL WORLDS

As Barton (17) says our aim is to study "problem complexity" not "algorithm complexity"; to evaluate complexity we need a tool easy to use and not so reductionist (18) to be misleading: it must be simpler than some mathematical models and more exhaustive than a single figure (pure number, time, negentropy or total information content). The comparison of the results given by RAM (Random Access Memory) and RASP (Random Access Stored Program Machine) made by Atlan show how is easy to get results difficult to evaluate (19).

The evaluation of the complexity of large technological systems may not considered by itself because it is strictly linked to the complexity of the artificial world, of the eco- and human systems. This brings a first problem because, as Atlan pointed out (20), it exist a substantial difference "between the complexity of artefacts and that of natural systems".

The technological systems may not be considered as closed and standing alone anymore; the connections among the three sectors are so strong that the steady increase of the level of complexity of the artificial world brings up a continuous increase in the complexity of the other two . The quantities of energy, matter and information flowing in and out of the artificial world are of such a magnitude to alter the natural cybernetic cycles that control the ecosystem stability ; the increase of the complexity level in the artificial system - as the one that could arise following the introduction of expert systems in industrial plants - involves the birth of new behaviors which requires a continuous improvement in the human knowledge to understand, design, control and govern the outer world.

This knowledge is not bounded to the physical systems but must be spread to other kind of systems; as Prigogine has written quite recently (21) in an economically devoted magazine "concepts comprised in the dynamics of complex systems offer new possibilities to cognitive sciences". Certainly a great conceptual gap has to be overcome to go from equilibrium , to the non equilibrium thermodynamics and from thermodynamic to logic ruled systems.

New behaviors of artificial world and of Large Systems.
The main causes of these behaviors may be found in new structural and operating aspects of modern artificial systems:
- the higher amplification capacity (or the higher gradients in energy fields)
- the increase of possible feed-back cycles due to the increasing interconnections,
- the morphogenetic and knowledge formation roles of information.

The "Large Systems" - which include artificial, human and eco systems - show very peculiar behaviors ; the regularities of these behaviors may be self-evident or, to be disclosed, require some previous knowledge based transformation. Anyway the presence of these regularities indicate the existence of an inner order formation capacity (22).

The first point is then to ascertain how this order making capacity is related to the complexity; the debate about the limits of the order formation capacity by non living systems has been very lively , we may remind the contributions by Schroedinger, Monod, Varela , Prigogine, Jantsch, Haken on: negentropy, casual and causal process, autopoiesis, self-organization, open system thermodynamics, synergetics ...

Today in physics the "order from chaos" phenomenon is receiving much attention ; the extrapolation of the results to wider domains is surely justified (Large Systems must obey the laws of physics) provided that due care be given to the changes that such an extrapolation requires. Special attention has to be given to the possible closure of feedback cycles through the flow of information or negentropy; i.e. : the energy distributed evenly on many degrees of freedom may be concentrated - ensuing a resonance or a logic choice - in a single one reaching a threshold value.

In cybernetics there have been some proposals to try to classify the systems and to correlate the behaviors to the kind of structure (table 1) and to the role of information: taking in account that the characterization of a system depends on its internal organization, on its logic , following the greeks we may assume that the type of the system is individuated by its "language" (= organized information) .

Table 1: ELEMENTARY SYSTEMS BEHAVIORS

A : Maruyama and Bush classification

Cybernetic Class	Behavior
O	passive
I	regulative
II	evolutive
III	human
IV	social

B: established on the base of the "language" (35)

Language	Behaviors
entropic	passive
automatic	regulative / evolutive
logic	anamorphic / intelligent
noetic	thoughtful / communicative

The problem in Large Systems has not a purely academic interest; we must know:

- what is the result of causal processes and what depends on the blind law of chance,

- which behavior must be attributed to the laws of physics and which one depends on the free will of man ,

The philosophical and theoretical aspects of the second point have been dealt with extensively for instance by Morin (23) but the practical implications in Large Systems design, control and the linkages with complexity are far from having received an exhaustive answer.

The government of the complexity of large technological systems requires a sound understanding of two main aspects:

- "auto" and "self" evolution
- knowledge formation

Self and autos

Many authors have dealt with such problems but on a system approach is necessary to avoid any risk of misunderstanding.

Semantically speaking autos are spontaneous process devoid of any subjectivity: any system with a negative resistance or a feedback may - when some technical conditions are satisfied - show an autonomous or spontaneous behavior; i.e. Prigogine and Haken theories are very well known and many experiments in fluidodynamics (Benard) , in chemistry (Zabotinsky) and in electromagnetics (laser oscillators) show impressively the possibility of an order formation from chaos.

Staying always in the domain of "Auto" systems , much less is known - due perhaps their novelty - of the order formation capabilities of purely logical systems; some studies have been done in cellular automata but we are far from the complexity level that can be reached in a plant where a computer may interact with the logics of the system. We have to consider that in so doing we are mooving from an intermediate complexity level - not far from the understanding capacity of man - to a level much higher that can be beyond the "von Neumann Threshold".

The "self" behavior presume a subjectivity , a free will, a decision capacity not only a logical deterministic choice; then it is easy to understand why the "self" behaviors are very often judged completely chaotic and unforeseeable . These conclusion are unacceptable if we assume that the man is a rational being ; "self" behavior may be very complex but not random . Perhaps it should be of some interest to make some comparison with the "deterministic chaos" which is receiving so much interest in physics.

Many psychologists or sociologists are dealing with such problems but it is difficult to master the complexity of self behaviors in a way that could give a contribution to the solution of the problems of Large Systems.

Too often the spontaneous evolution of relatively simple auto-systems is attributed to a high level of complexity when the only reason is that the threshold of higher mode evolution has been overcome ; in economic and social sciences the same kind of error is made when the multimodal oscillations of the systems are attributed to complexity instead of the simple increase of dimensions above some threshold value. The confusion may arise by the fact that some scientists (24, 25..) consider "self-organization" referring to an increase of complexity due to physical process (in the sense of Caianiello) in the frame of an already existing organization.

In large plant this is one of the main reasons for very serious pitfalls. If a system is showing a "auto" behavior its resilience to any perturbation may be very high and the only way to interfere may be to change the "feed-back" condition (analogical signal or logic choices): which is a counterintuitive action; if , instead, we have to deal with a "self-behavior" we have to intervene on the knowledge formation system or on the basic criteria of judgement. The first action may be relatively simple , the second one could be very complex.

These aspects are very important in Large Scale Plants because the change from a purely hyerarchical structure to a modular one pose the problem of the degree of freedom and judgement in the modules.

Knowledge formation

The increase in the complexity of Large Plants may be mastered only by an increase in the understanding of the inner rules and laws of the system. This requires to get a better knowledge not only of the technical aspects but also ·of the more theoretical ones. Operators and designers are subject to more deeper education training but man has inherent limits in: sensorial perception, speed of information processing, degree of alert, standard of performance, resilience to external physical actions, memory access, … . Man can successfully handle only problems of a limited - intermediate - complexity ; the solution for higher complexity systems is to resort to groups of men or of men-machines : in this way we try to match the complexity of the plant with the complexity of an organization. This introduce the necessity to fully understand the behavior of these kind of homogeneous or dishomogeneous groups; the problem is not so easy today and it will become more difficult in the future because also the introduction of interconnected and parallel computer brings the complexity in a new light.

Expert systems are no more evaluated on the "data base" capacity but on the Knowledge acquisition and representation capabilities; the evaluation of this aspect is certainly not so easy and it poses very serious basic problems on complexity evaluation because as D.Sahal says: "one prominent attribute of cognitive structures is complexity" (26) .

Human engineering on these days may progresses very rapidly, the better understanding of the human mind allows the designer of plants to better define the capabilities and limits of human operators; one point is receiving a lot of attention : the behavior of man under stress or strong emotions. One of the latest solution is not to rely on an operator emotionally stressed; the design of the plants are conceived in such a way - called "walking away operator" - as to leave the operator the time to recover : the reason of this choice is due to the assumption that intuitions and emotions are not very helpful in a highly complex artificial system with a counterintuitive behavior.

Algorithmic, structural , design, control complexity concepts should be revised in a unique framework; such a revision may only start from the basic fundamental sciences like physics or mathematics. Surely a problem will be the transition from micro to macro world : the change from "the degree of freedoms" to the "variety".

A few examples

We have seen the reasons why the definition and the measure of complexity are of paramount importance for design, control and operation. May be it is time to give a few examples of what does it mean for Large Systems which are made up of different kind of apparatus and subsystems.

Uranium enrichment plants

To start with we shall discuss an uranium enrichment plant, not just because it is one of the largest technological systems, but also because it is a good example of an order formation system (it separates isotopes).

The plant structure, and its complexity, change with the kind of technology adopted; it is worthwhile to mention that the intuitive evaluation of complexity, as practical experience has demonstrated, it has been completely wrong: huge plants with millions of separating elements are simpler - to design and to operate - than others smaller plants with only a few thousands pieces.

An industrial enrichment plant, based on gaseous diffusion process, contains over 10 millions of so called barriers. Physicist and engineers, fully aware of the problem that such aspect could arise, have found ways to reduce the apparent complexity of the plant: the first ones have developed the theory of the "ideal cascade" , the seconds ones have assembled - clustered - the separating elements in stages and "square cascades" in such a way as to approximate the "ideal cascade" to a very high degree. As a result it is possible to evaluate the behavior and the performances of a large plant with very simple formulas. It has to be mentioned that the theory of the "value function", the key point of all the problem, has been elaborated by Dirac.

When other more advanced technologies - like centrifuges, laser ionization or molecular excitation - were proposed in a first instance it had been supposed that to reduce the number of separating elements - respectively a few thousands and a few tens - it should have reduced the problems of the complexity of the system. But it was soon evident that this was not the case; the separating elements had much higher design and control complexities and their smaller number did no anymore justify any transitions to the continuum and the discretization of the plant - that had to be maintained - introduced a new kind of problem, like recycling and mixing losses.

In laser isotope separation plants the description level has to be kept below to the atomic or molecular levels : the so called red shift was a crucial point for the UF^6 molecular process.

With centrifuges it was also difficult to get the same stability of gaseous diffusion plants where the barriers - highly dissipative and disjunctive elements - quench any small scale perturbation and raise the time constant of large perturbations to values so long (months) that the control problems could be easily solved in an "anticipatory" manner with the aid of low level complexity computer.

From this kind of plant we than can learn three lessons:

- *the complexity is not strictly linked to the number of elements,*

- *any plant has its own elementary structural dimensions that fix the scale on which the design complexity has to be evaluated,*

- the "geometric" scale for the static design may be completely different from the "time" scale that has to be adopted to evaluate the control complexity.

The last point is very important because it happens that the complexity of the plant, forgotten in the simplification of the static design, deploys its full dimensions when dynamic problems are considered ; i.e.: the short time scale raises the difficulty to get what top experts in cybernetics (27) call an "anticipatory" control.

Nuclear power plants

The nuclear power plants are interesting from a complexity evaluation point of view because they are made of a relatively few - tens or thousands - main items of a very different type. Referring to the classification proposed in table 1 we can see that the reactor itself is an auto-system: the nuclear chain reaction is said (improperly) to be self sustaining and posses an evolutionary capacity.

Many of the problem of control of the plant are linked to the necessity to govern the evolutionary behaviors of the reactor ; these behaviors for some kind of technologies are more complex : multi mode spatial or xenon oscillations; temperature. pressure or power negative reactivity coefficients. For these reactors the complexity is so high to overcome the capacity of a single operator and provisions are made to help him to perform his duties : automatic controls, safety systems, emergency shut down systems... Anyway in the western countries some reactor types were discarded due to the too high control complexity.

Notwithstanding all the effort the evolution of nuclear power plants, as practically all large industrial plants, has been characterized by a continuous and steady increase in complexity. In a meeting of U.S. utilities organized in 1982 by the EPRI (Electric Power Research Institute) the general feeling in the United States , especially due to local conditions and federal organization rules, was that the autocomplexification phenomena had gone too far. In 1985 A.M. Weinberg and I. Spiewak , claiming to show a route for "a new start for nuclear power" pointed out the necessity to counter the autocomplexification tendency and look for simpler solution (28,29). The solutions being proposed - modular structures, passive or inherent safety features -certainly require a very keen evaluation of structure and operators complexities .

The autocomplexification

The problem of "autocomplexification" of industrial plants and products, - an aspect to which it would be noteworthy to devote a great deal of attention - it is bringing a lot of systems up to a high degree of complexity. On this point Ashby (30) noticed that computers have given the man the opportunity to handle systems of an intermediate degree of complexity : enough complexes to be interesting and enough simples to be easily understood.

The complexity increase of new technologies (31) and of their products (32) poses very serious problems not only on the technological field ; if a threshold of complexity is overcome the Large Systems show very well defined autocomplexification evolutions : increase in number of components, procedures, rules , personnel, data handling, paper handling ... . This autocomplexification may bring to open , widespread, subsystems or end with the formation of parasitic systems closed to information flow. The evaluation of the threshold values of complexity for the two different kind of autocomplexification is of a paramount importance.

The butterfly effect

The stability and the kind of evolution of technologies (of their markets and plants) is very often linked to small - or even inappreciable - differences of respective complexities ; in complex physics systems - such as the meteorological, gasdinamical ,...ones - the so called "butterfly effect" is very well known.

Than the degree of approximation of the complexity measure has to be such to avoid a complete misunderstanding of the situation: the shift from deterministic to chaotic behaviors may not be easily appreciated. If a reliable quantitative measure is not available, we shall have to rely on qualitative evaluation.

3. DIFFERENT KINDS OF COMPLEXITY

Different meanings of complexity

D. Sahal reminds (33) that "one of the first attempt to measure complexity was made by Wrinch and Jeffreys in 1921 in a study of certain fundamental principles of scientific inquiry".

The complexity is a subjective quality, its meaning and its value, change following the scope of the system being taken under consideration. D.Sahal (ibidem) makes a distinction between ontological (that of material objects) and semiotic (that of ideal objects) complexities and he quotes 4 dimensions of the semiotic one.

Design , control or structural complexity normally are not the same; as Kolmogorov pointed out "while a map yelds a considerable amount of information about a region of the earth's surface, the microstructure of the paper and the ink on the paper have no relation to the microstructure of the area shown on the map" (34).

One of the first scientists to deal with the problem of complexity was certainly Weaver (35) who in the 1948 raised the problem of organizational complexity. It is impossible here to quote all the persons who have dealt with the different aspect of complexity in the different domains. Perhaps it is worthwhile to remind that two of the more well known experts on control complexity and information theory, Yudin and Goryashko (36), are of the opinion that the increase in complexity requirements - only for the control of systems - is much more than the increase in complexity capacity of computers (37).

These opinions may certainly be shared by who has followed the development of Artificial Intelligence (38) : the introduction of connectionism and parallel computing may be considered as a tentative to increase of order of magnitude the structural complexity of computers: well beyond the limits achievable only with the increase of speed of computing and memory capacity. Today the memory of the Vth generation computers is no more considered as a storage of data but as an assembly of knowledge; this is completely reversing the past trends (39) but it requires a new look to the complexity problems based on knowledge formation.

In the computers of the future - which will have to interact with real systems - it will be necessary to find more exhaustive definitions for design , computational and algorithmic complexity (compatible with plant complexity definitions); it is difficult to think that these kinds of complexity could be evaluated only by a number or by a computing time.

It looks to me now that the main problem perhaps is to try to unify the different aspects and approaches but that no immediate solution is available (40).

Complexity measures in small and large systems

In the already quoted paper D. Sahal states : "the argument must be rejected that complexity is too elusive for measurement" ..."if appropriately derived the concept of *negentropy* as a measure of complexity will suffice for most application." The second part of this statement presents some problems when the item of which we want to measure the complexity is a "Large Heterogeneous System"; to apply this kind of evaluation is necessary to lose all the multiform variety of the system . We must remind that for the control of complex systems Ashby considered the variety as the first requisite. The thought of Ashby has been often misunderstood and may be useful to compare his thought to the one of Wiener. In military systems the latter was concerned with the error reduction of a antiaircraft gun, the main concern of the former looks like it was the north-atlantic submarine war : the main point was to have ships and weapons suited to the enemy warships. We must remind that in the Pacific war, japanese navy lost a very famous battle because the planes on the aircraft-carrier were charged with bombs instead that with torpedoes.

The difference between large and small system complexity is the same that exists between strategy and tactics. For large system we may not avoid "variety".

4. EVALUATION OF LARGE SYSTEMS COMPLEXITY

A definition and its consequences

To evaluate the complexity of Large Technological Systems we may start from the following definition:

A system is complex when it is built up of a plurality of interacting elements, of a variety of kinds, in such a way that in the holistic result no evidence can be traced of the characteristics of the singles elements.(41)

This definition encompasses the one by Rosen that says : *complex systems are the ones to which the newtonian paradigma does not apply* .

The values of the plurality and variety - which constitute a necessary condition - establish the level of complexity. We have to take in account that in a Large System many kind of "languages" are being used (here *language* means organized information of any kind).

The complexity has not an absolute value but a relative one which depends on the choice of the definition level or of the minimal dimensions of the constituent elements.

The choice of the level is influenced, besides subjective judgement , by objective aspects like : interaction levels, degree of freedom, logical or organizational links.

If the definition is pushed to the extreme limit we should have to consider , following the greek philosopher, the unbreakable elements (αθΩμ) of matter, energy or information. The variety of the world depends on the different complexity organization of the elements and this point out the great role of "variety" in complex system as was proposed by W.R. Ashby.

If this definition , apparently quite "naive" , is accepted , we have to consider that some very relevant consequences directly follow up .

First: if the components of a complex system have to interact, they must be "open" ; it follows that their behaviors are far different from the ones of closed systems that are often considered as usual.

Second: if the system is open - and presumably non linear as real systems usually are - is not possible to apply the newtonian paradigm following which categories of causation are essentially segregated in separate packages (42).

Third : the capacity of spontaneous evolution of complex systems is not linked to the level of complexity but to the nature of the elementary systems that, being obliged to stay open, show their evolutionary capabilities; on this point we may remind that Von Neumann (43) thought of the existence of a threshold of complexity above which entirely new models of behavior appear (such as self-reproduction, evaluation and free will).

The 4 attributes of complexity
Taking in account these aspects and reminding the theories of Ashby (on the complex systems control and the roles of information), of Sahal (for simpler systems) of Casti (on connectivity) - that is not possible to summarize here - one may propose to evaluate the complexity of systems by the help of 4 attributes :

- *Numerosity* : total number of elementary components
- *Variety*: number of different kind of elementary
  systems
- *Type* of components
- *Organization*

To measure these attributes one could deal separately with the main aspects of the system : statistics, cybernetics, information theory, connectivity ....

The measure of these attributes is not always very easy and poses the problem if it is better to have a meaningful qualitative appreciation or a very exact numerical measure devoid of any physical meaning. Surely as lord Kelvin was saying if there is no measure there is no science, but given the different applications of the complexity concept, some of these attributes could be evaluated only in a very definite context. However we must remember that complexity is always a subjective and relative quality : its evaluation depends on the purpose of the "observer" - that can be any kind of outer (or even inner) system - and on the degree of description or analysis required. No radar alert station would tell the control center that $2 * 10^{26}$ atoms are approaching at 900 miles/ hour !

This is the reason why we have spoken of "language" in the definition of elementary systems : when the language has been established we know the meaning of the nouns. This aspect is very important in the case of Knowledge based AI systems : any noun has its own information content and then its internal complexity . This complexity could or not be taken in consideration according to the description level required.

Numerosity

Surely there is no major problem for numerosity; its measure is a pure number and then - apart problems connected with arithmetic or computational complexity and computer capacity - no further analysis is required when the level of description has been established.

When numerosity is very high we may presume that statistical aspects could be dealt with in its framework; this is the reason why sometime we try to aggregate the same elements forming clusters or making larger size elements.

In the Large Plants these aspects are very well known because the level of description is different when static or dynamic problems are considered; in the second instance, the necessity to perform the calculations in a time congruent with the times constants of the systems, it requires that the description level be fixed to a value much higher than in the first one where computing time are not so limited.

Variety,

Also this attribute is measured by a number and being normally referred to a much smaller population, has a less aleatory character. Also in this case it is necessary to subjectively establish how equal has to be two items to be considered of the same type . Before the nuclear age two atoms of the same element where considered of the same type and no one was caring about isotopes!

The color of a car has no discriminatory value apart from the factory painting (once upon a time Ford or Volkswagen cars were all black) or accident propensity statistics (bright color are safer) are considered.

Type

As a general rule we have assumed that a "type" is defined by a "unity of language" . The problem is what kind of unity of measure may we adopt ?

The definition of the type depends completely on the language which has been assumed : its semantic value. When we say that a microprocessor is a 80286 we mean a item perfectly defined in computer language: its complexities in other languages - structural, chemical, solid state physics, ... are well established and could be called upon when desired.

We may think that a "type" - specified by a noun , a formula, a symbol, a icone - is corresponding , in the language being chosen , to a model: the validity of this modelling procedure will be limited to the aspects under consideration : design, control, thermodynamics, cybernetics, computational, knowledge representation ...

The level of description and the unity of measure(s) - number, time, negentropy, information , degree of freedom, shape ... - will change case by case.

In any instance it will be possible to evaluate the complexity of the model because its level of description is established a priori The complexity of a landscape in nature is very difficult to be defined; but, when its model - a picture, a tv image - is considered, it is relatively easy to establish the upper level of complexity; the same should apply to a control system : its model, if specified with a standard language, will have a very well defined complexity level.

This is not a novelty in the field of the evaluation of mathematical complexities where many theoretical discussions have been done on the typical algorithm or on the standard computer to be used to evaluate the complexity. Today in the domain of AI this kind of problem is very well known in relation to the knowledge acquisition and representation; an expert system of tomorrow has to know what is the meaning of a certain type of patient well beyond the logical noun.

Organization

To measure "organization" is very difficult because we may have very different kind of organizing structures: scientists (Simon, Casti, Caianiello, ...) normally are giving their main attention to the hyerarchical one which may be very well defined. Also for this one, when we start to consider the partition phenomena, the situation is no more so simple. Casti propose (op.cit. p.36) to use D-graphs for "connectivity" evaluation , but this kind of formalism apply only to some kind of organization; R.C. Conant (op. cit.) try to make a distinction between strong and weak interactions ; Atlan in 1974 published a Formal Definition of Organization. Up to know does not look as a general consensus has been reached on what kind of measure to adopt.

This attribute can be evaluated on a knowledge based language; in this case the difficulties are greater than in the previous case because we are generally lacking of a language that can describe in detail the different form of organization; Gottinger (op.cit. p. 184) enumerates a few classical examples of connection : serial, serial parallel, cascade, cross. Anyone who has worked on a separating plant knows how many types of cascade are possible to design with very different characteristics.

The most crude measure that can be proposed if the chief interest is thermodynamical or informational is : efficiency or noise production. This is normal practice in very complex technological systems; no one is caring very much about what happens in the details: what matter is what losses - in energy, matter or information - the organization introduces. In this view the elementary components produce and the best that the organization can do is not to waste the single efforts. This approach is the one adopted in isotope separating plants where the cascade efficiency may be increased by design improvements from 70 - 80 % to over 90% .

This approach is completely inadequate when we have to deal with very sophisticated organizations which do not limit to transfer - information, matter or energy - between the different items in a purely passive way but perform some kind of operation passive or active. One solution is to introduce fictitious elements to act and leave the organization completely passive made up only of "trasmission channels". The problem is that also in such a way we may not easily grab the "logics" of the organization : the wholesome project. When we deal with an artefact we may presume to know the purpose of the designer and from it we may deduce the project. The problem is when we have to deal with nature made items : the first question , never exhaustively answered, is : "does it exists a purpose?"

When we consider large multipurpose plants this question becomes a very serious and practical one; the time when the only purpose was to produce cheaply are far gone . Today we no more speak of a simple - one valuefunction - optimization but we have to strike a balance between many opposite requirements: economic performance, quality of life, safety, national strategies, technologies substitutions, industrial development. So complex problems have been dealt by manager organizations mainly on a practical experience base; to day the complexity has reached so high levels that a more scientific approach should be highly desirable.

In a static view it is possible to consider the degree of freedom that the organization can control: if this value is equal to the total sum of the degree of freedom of individual components and if no variety existed , the system as a total should be completely controlled or enslaved. Anyway if, on the contrary, the variety was different from zero the system could present some degree of freedom because the organization could not fit perfectly to the system. This is what happen for instance in the ability games where the grab and the object geometry are different : triangular versus square.

Certainly all these measure are very crude and misleading because they do not offer any mean to evaluate the more complex phenomena linked to the organization : intersystem looping, cross-talks, crossmodulation, linkage splitting or resolution, lock-in , cluster formation, mutations ...

These aspects should not be forgotten in Large Systems; everyday we see what happens when they are underestimated. The autocomplexification phenomena are certainly strictly linked to these kind of events. In a large system - a big industrial plant, a town... - the free space is always diminishing (a consequence of the well known territorial expansion), the communication and traffic networks jams, groups of individuals closed to any outside influence are being born, ...

5.CONCLUSIONS

The measures of complexity are a subject of the uppermost interest not only by an academic point of view; the steady increase in complexity of industrial plants and products poses a serie of a new kind of problems in the fields of structure design , control, safety, government, planning .

The solution of these problem require a very keen measure and comparison of different kind of complexity and may be found only .in a multidisciplinary approach that could solve not only the problems inside the plant but also the ones linking the plant to the environment and to the human social system.

The deep knowledge of the problem of complexity acquired in the scientific domain must be transferred as soon as possible to the technological and social domains to help for solving very urgent problems on spontaneous behaviors, stability and government. In these points of view the measures of complexity should be not only scientifically sound but also easy to use by laymen ; if this will not be possible it should be helpful to have well defined: concepts, qualitative evaluation rules and procedures. This could gives a great contribution to avoid mistakes and misunderstandings.

For engineering purposes, my suggestion is to evaluate separately 4 different aspects of complexity : *numerosity, variety, components type, organization* ; if this approach is going to be considered acceptable a considerable theoretical effort should be required to formalize the measurement of these aspects for the different types of complexity.

1 S. AIDA,et al., *Science and Praxis of Complexity*, United Nations university 1985
2 H. ATLAN et al. *La sfida della complessità* , Feltrinelli 1985
3 G. PASQUINO, *La complessità della politica*, Laterza 1985
4 D. WARSH, *The Idea of economic complexity* , Viking Press 1984
5 H. GOTTINGER, *Coping with complexity* , Reidel, 1983,
6 J. CASTI, *Connectivity, complexity and catastrophe in large scale systems* - IIASA , Wiley 1979
7 K. BOULDING , *Learning by simplifying complexity: how to turn data into knowledge* , in "science and praxis of complexity", editor S.AIDA
8 J. CASTI, *Connectivity, complexity and catastrophe in large scale systems* - IIASA , Wiley 1979
9 G. KARPMAN, B. DUBUISSON, *Complexity Index for System Diagnosability* , IEEE Trans. SMC, vol.smc-15,n.2 , march/april 1985
10 R.U. AYRES, *Complexity, reliability and design: the coming monolithic revolution in manufacturing*, IIASA WP-86-48
11 R. ROSEN, *On information and complexity*, IIASA CP-85-19
12 R. ROSEN, *The physics of complexity, Ashby memorial lecture*, in : Power, autonomy, utopia, by R. Trapple , Plenum press, 1986
13 J. RASMUSSEN, *The role of hyerarchical knowledge representation in decision-making and system management*, TRANS IEEE - SMC - vol. smc-15,n.2, march/ April 1985
14 E.W. PACKEL, J.F. TRAUB, *Information-based complexity*, Nature, vol. 328, 2, july 1987
15 D. BLACK, J. MANLEY, *A logic- based architecture for knowledge management*, IJCAI 87 , Milan 23-28 Aug., pp 87-90

16 E.R. CAIANIELLO, *Topics in the general theory of structures*, D. REIDEL, 1987

17 G.E. BARTON, R.C. BERWICK, E.S. RISTAD, *Computational complexity and natural language* , Bradford Book , MIT Press,1987.

18 K. BOULDING , *Learning by simplifying complexity: how to turn data into knowledge* , in "science and praxis of complexity" , editor S.AIDA p 25.

19 H. ATLAN, *Natural complexity and self creation of meaning*, in "Science and Praxis of complexity" , editor S.AIDA, p. 175

20 ibidem p.173

21 I. PRIGOGINE , *Una nuova razionalità?* , Il pensiero economico moderno , anno VII, gennaio-giugno 1987 - n. 1-2.

22 For further information see : IIASA reports of C. Marchetti, books of Devendra Sahal, Technological Forecasting and Social Change Journal

23 E. MORIN, *Self and Autos* , in Autopoiesis , M. Zeleny editor, North Holland 1980

24 E.R. CAIANIELLO op. cit. p. 8

25 H. ATLAN, op.cit. p.179

26 D. SAHAL *Elements of an emerging theory of complexity per se* Cybernetica, vol. 19, 1976, pp. 5-38.

27 R. ROSEN, *Anticipatory systems* , Pergamon 1985

28 A. M. WEINBERG, I. SPIEWAK , et al. *The second nuclear era,* Praeger, 1985

4 A. M. WEINBERG, I. SPIEWAK , *Inherently safe reactors*, Ann. Rev. energy 1985, 10 : 431-462

30 W.R. ASHBY, *An introduction to cybernetics*, Chapman & Hall 1956

31 M. ZELENY , *La gestione a tecnologia superiore e la gestione della tecnologia superiore,* in "La Sfida della Complessità" , Feltrinelli 1986

32 L.U. BUSINARO, *Dall' ameba all' automobile,* 1987

33 D. SAHAL *Elements of an emerging theory of complexity per se* Cybernetica, vol. 19, 1976, pp. 5-38.

34 A. N. KOLMOGOROV , *Three approaches to the quantitative definition of information* , Problemy pederachi informatsii, vol.1, pp 3-11, 1965

35 W. WEAVER, *Science and complexity* , American scientist , vol 35, oct. 1948, pp 536-545

36 D.B. YUDIN, A.P. GORYASHKO, *Control problem and complexity theory*,P.I, II, III, Eng. Cyb. vol 12, 13, 14; 1974, 1975, 1976

37 G.B. SCURICINI , M.L. SCURICINI, *CIBERNETICA E NOETICA* , Sansoni, 1985, p.107.

38 10 th IJCAI Conference Milan 22-28 August 1987

39 G.B. SCURICINI op.cit. p. 236

40 G.B. SCURICINI, *La complessità* , un tentativo di approccio sistemico, Conf. ENEA, 8-9-87.

41 G.B. SCURICINI, ibidem

42 R. ROSEN ,1985 op.cit. p.18

43 H. GOTTINGER, *Coping with complexity* , Reidel, 1983, p. 12

AN INTRODUCTION TO THE THEORY OF COMPUTATIONAL COMPLEXITY

D.P. Bovet and P.L. Crescenzi
Dept. of Mathematics - University of Rome "La Sapienza" - Italy

0. INTRODUCTION

In this introductory paper, an attempt is made to present the key concepts of the theory of computational complexity and to review some of the most interesting results obtained in the last decade.

The paper is organized in two parts: in the first 4 sections, the theoretical framework of computational complexity is introduced and a formal definition of problem is stated, together with that of dynamic measure of complexity. In sections 5 and 6, some recent applications of that theory to parallel computers and to artificial intelligence are presented.

1. WHAT DOES COMPUTING IMPLY

The notion of computing is based on a few fundamental concepts. First of all, computing is done referring to a specific *computation model*. Next, computing consists of executing a sequence of steps on the computation model as specified by a suitable *algorithm*. Finally, computing refers to some *input data* upon which the algorithm will operate.

As an example, a computer is a computation model, a program written in a programming language recognized by that computer is an algorithm, and a file of data to be read by the program are input data.

Computational complexity, as an abstract discipline, does not refer to specific details about computers like instruction execution time or memory cycle. The main objective is to enucleate the intrinsic complexity of some problems, or the goodness of some algorithms, without being tied by existing architectural differences among computers. The underlying assumption, which has been confirmed in many cases, is that if a problem is really hard (or if an algorithm is realy good), this will show up on any type of computer.

2. A WELL KNOWN COMPUTATION MODEL

As a consequence, a rather abstract computation model which embodies the key features of today computers is used. The model called *Turing machine* (TM)

was introduced in [8]: physically, a TM consists of a Central Processor Unit and a read-write tape. The tape is divided into an infinite set of cells and the CPU can read, at each time, just one cell, called the <u>scanned cell</u>, (see fig. 1). The symbols written on the tape belong to an alphabet Σ. The CPU reads the symbol σ contained in the scanned cell, and, depending on σ and its internal state, it performs one or more of the following actions:
   - write a new symbol in the scanned cell;
   - move to the right or to the left;
   - switch from one internal state to another.

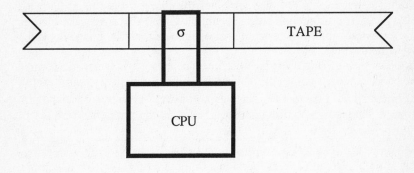

Fig.1 A Turing machine

According to the previous definiton, an *algorithm* for a TM is a set of quintuples of the form:$<q_i, \sigma_j, \sigma_l, m, q_k>$, where q_i and q_k are two internal states of the CPU, σ_j and σ_l are two symbols of Σ, and m can be r(ight), l(eft) or s(tay). The meaning of each quintuple is the following: if the CPU is in the state q_i and it's reading σ_j, then it writes σ_l, performs the movement specified by m and changes its state from q_i to q_k.

At the beginning of a computation, the tape contains the <u>input</u>; every performance of a quintuple is called a <u>step</u> of the computation. The CPU, finally, has a special state, called <u>final state</u>: when it reaches this state, it stops the computation; at the end of the computation, the content of the tape is called the <u>output</u>.

Several variations of the standard TM have been introduced in the literature (TM's with k tapes or with k heads, TM's multidimensional, nondeterministic TM's, probabilistic TM's, etc.), and the relationships between the different variants have been studied (see [5]). In this context, however, those variants will not be considered. The main reason is that, although the standard TM is a very simple model, it can be conjectured that every "computable" function can be computed by a TM (a function f is computed by a TM if, given in input x, it stops with output f(x)). This conjecture is known as *Church's thesis*. In fact, nobody has been able to find a computation model more powerful than TM, and it would be very surprising if somebody succedeed in doing so.

3. LANGUAGES AND DECISIONAL PROBLEMS

Given an alphabet Σ, let Σ^* denote the infinite set of all possible strings of elements of Σ. A *language* L on Σ is a subset of Σ^*. A language L is said to be *recognizable* if there exists a TM that computes the characteristic function of L. It can be proved, using diagonalization techniques, that there exist languages which are not recognizable; however, the set of recognizable languages is large enough to be "interesting".

In fact, the notion of language is introduced because it is a natural formal counterpart of the intuitive notion of decisional problem, which is central in this discussion. A *decisional problem* Π is a "yes-no answer" problem, consisting of a set I_Π of *input instances* each of which specifies what the problem structure is (e.g. a graph, a boolean formula, a mathematical equation, etc.); and of a *question* about the instance which admits only a yes or no answer (e.g. does the graph admits a path of length k?, is there an assignment of values for which the formula assumes the value true?, is there an integer solution to the equation?, etc).

To associate a language on an alphabet Σ with a decisional problem Π, the first step is to encode every input instance of Π into a string of Σ^*. In other words, an *encoding function* $\chi: I_\Pi \to \Sigma^*$ which partitions Σ^* into:

- the set of strings that encode the instances of Π which have answer yes;
- the set of strings that encode instances of Π which have answer no;
- the set of strings that don't encode instances of Π.

The first set is called the *language associated with Π* under the encoding χ. A problem Π is *decidable* if the associated language is recognizable. Complexity theory deals only with decidable problems.

For sake of simplicity, a problem Π will be identified with its associated language; thus, an input instance of Π is a string x of Σ^* and solving a problem consists of deciding if x belongs to the language associated to Π. Finally, the length of the input instance x, that is the number of cells necessary to write x on the tape will be denoted as |x| .

4. THE COMPUTING COMPLEXITY OF DECISIONAL PROBLEMS

In order to define the complexity of problems, a computing complexity measure is needed. Two different kinds of measure called static and dynamic have been proposed.

A *static measure*, simply speaking, depends only on the program used to solve the problem, but doesn't distinguish between different instances of the problem itself (the number of quintuples of a TM is an example of static measure).

A *dynamic measure* is related both to the length of the input instance x and to the computation having x as input. Although static measures have some very interesting properties, this tutorial paper will focus on the dynamic ones.

As shown in [1], any dynamic measure must satisfy the following two requirements:
- the measure is defined if and only if the computation stops;
- it must be always possible to decide if the measure is less than or equal to a fixed number.

Two natural dynamic measure are the number of steps performed by T during the computation with input x, in symbols TIME(|x|), and the number of cells scanned by T during the computation with input x, in symbols SPACE(|x|). It is easy to verify that those measures satisfy the two previous requirements.

It is now possible to define the computing complexity of a decisional problem. Problem Π has *lower bound* f, if any TM that solves Π has time complexity TIME $\in \Omega$ (f) (i.e. there exist two constants k_1 and k_2 such that TIME(|x|) $\geq k_1 f(x) + k_2$). For example, sorting n numbers in increasing order requires TIME(n) $\in \Omega$ (nlogn). Problem Π has *upper bound* g, if there exists a TM that solves Π with time complexity TIME $\in O(g)$ (i.e. there exist two constants k_1 and k_2 such that TIME(|x|) $\leq k_1 g(x) + k_2$). For example, it is possible to perform the multiplicationof two n×n matrices with TIME(n) $\in O(n^2)$.

From the previous definitions, an important question arises about the encoding function χ introduced in the last section. This function must be, in a certain sense, "reasonable": in other words, it must be concise, because a prolix encoding would allow any problem to have a very "low" upper bound; and it must be also essential, in the sense that every part of the encoding string must be used to solve the problem. With this idea of reasonable encoding in mind, it can certainly be asserted that every problem has a linear lower bound.

Obviously, the intrinsic computing complexity of a problem can be considered fully investigated only when it has been possible to derive lower and upper bounds close to each other. Unfortunately, this does not happen very often and the known bounds are not very tight. Figure 2 represents a typical situation.

"Easy" problems can now be characterized in terms of their upper and lower bounds. Formally, a problem is called *tractable* if its upper bound is n^k for some constant k. For example, the problem of finding the shortest path in a graph is tractable. A problem is called *intractable* if its lower bound is k^n for some constant k. Proving the intractability of a problem is much more difficult than proving its tractability; most of the known intractable problems are, in fact, "conjectured" to be intractable, but nobody has been able, till now, to give a formal proof of their intractability. For example, the problem of deciding whether a given boolean formula is satisfiable is conjectured to be intractable.

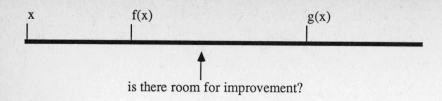

is there room for improvement?

Fig.2 A typical situation for the bounds of a problem

Figure 3, taken from [3], clarifies the distinction between tractable and intractable problems: n represents the length of the input and it is assumed that the problem is run on an hypothetical computer executing one step each microsecond.

n \ TIME	10	20	30	40	50	60
n^2	0.0001 seconds	0.0004 seconds	0.0009 seconds	0.0016 seconds	0.0025 seconds	0.0036 seconds
n^5	0.1 seconds	3.2 seconds	24.3 seconds	1.7 minutes	5.2 minutes	13.0 minutes
2^n	0.001 seconds	1.0 seconds	17.9 minutes	12.7 days	35.7 years	366 centuries
3^n	0.059 seconds	58.0 minutes	6.5 years	3855 centuries	2×10^8 centuries	1.3×10^{13} centuries

Fig.3 The difference between tractable and intractable problems

5. THE COMPLEXITY OF PARALLEL COMPUTING

Let us consider the following problem: n persons p_1, p_2,..., p_n must find the maximum of n numbers a_1, a_2,..., a_n. If the n persons cannot comunicate with each other, then any program that solves this problem has lower bound $\Omega(n)$: in fact, each person must examine all the numbers to find the maximum. Suppose now that each person can comunicate with the others by means of a shared memory. Then the following simple program will solve the problem with upper bound $O(\log n)$ (for sake of simplicity, we suppose that $n=2^k$, for some k, and that each person can find the maximum of two numbers in one step):

step 1: let p_i copy a_i in the cell X_i of the memory;

step 2: for each i, $1 \le i \le n/2$, let p_i find the maximum between X_{2i-1} and X_{2i}, and let he/she write the result in the cell X_i of the memory;

step 3: set $n=n/2$;

step 4: if $n>1$ then go to step 2 else stop.

At the end of the computation, the maximum will be in the cell X_1 of the memory. It is easy to verify the upper bound of the previous program.

This example shows that if several CPUs working on a common random access memory are available, then there exist problems called *advantageously parallelizable* which can be solved much faster than with only one CPU. At the same time, it is easy to verify that not all programs run faster on parallel computers: the goal of the computational complexity theory is to characterize advantageously parallelizable problems.

According to [4], a parallel computer is made of a CPU, a set of parallel processing units (PPU) and a common random-access memory arranged as shown in the figure 4.

The CPU contains the program of the parallel computer and it controls the behavior of the PPU's. Since the PPU's work in parallel, all the steps performed by the PPU's at the same time will be considered as a single step. Several parallel computer models have been introduced during the last ten years (see the excellent survey in [2]): they differ mostly on the way to solve possible write-conflicts between two or more PPU's (see [6]). The following results hold anyway for most of these models.

The notion of advantageously parallelizable problem is now formalized. If several PPUs are used to solve a tractable problem, then its upper bound will be expected to improve considerably. More specifically, a tractable problem is said to be *advantageously parallelizable* (a.p.) if there exists a parallel computer with a number of PPU's polynomial in the length of the input that solves it with $TIME(n) \in O(\log^k n)$, for some integer k. Thus, it follows from the previous discussion that finding the maximum among n numbers is an a.p. problem.

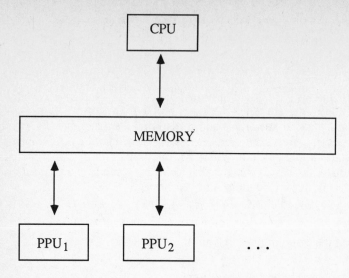

Fig. 4 A parallel computer

It seems likely that not all tractable problems are a.p.: although there is no formal proof, it is conjectured that several problems like linear programming problem and maximum network flow are not a.p.In conclusion, it has been possible to classify tractable problems into a.p. and not a.p. as shown in figure 5.

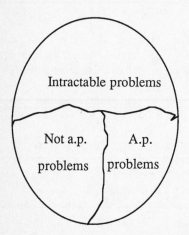

Fig. 5 A simple view of the world of decisional problems

6. COMPLEXITY THEORY AND ARTIFICIAL INTELLIGENCE

One of the main techniques of artificial intelligence consists of encoding non trivial problems into the language of mathematical logic and, then, solving them running an automatic theorem-prover.

Unfortunately, with a few valuable exceptions [7], very little attention has been paid to the inherent complexity required by such an approach. As a consequence, huge investments are made on so-called "expert systems" without any a priori information on the final level of performance of the product.

Researchers in complexity theory have investigated this problem using a worst case analysis approach.

The results presented in this section have been developed in [9]: they show that, in some critical cases, the complexity of automatic theorem-proving may turn out to be intractable even using the fastest supercomputers available.

Let $X = \{x_1, x_2,...\}$ denote an infinite set of *boolean variables*. A *literal* is either a variable x_i or its negation $\neg x_i$. A *truth-value assignment* (tva) is an assignment of the values **true** or **false** to any variable in X. If a tva assigns the value **true** (**false**) to the variable x_i, the literal x_i ($\neg x_i$) is said to be true under that tva and the literal $\neg x_i$ (x_i) is said to be false under that tva. A *clause* is a set of literals: it must be considered as a disjunction of literals, in the sense that a clause is true under a tva iff one literal in the clause is true under that tva. The empty clause Λ is false for any tva. A *formula* is a set of clauses: it must be considered as a conjunction of clauses in the sense that a formula is true under a tva iff any clause in the formula is true under that tva. A formula is said to be *satisfiable* iff there exists a tva under which the formula is true. Otherwise the formula is said to be *unsatisfiable*.

The following problem is now considered: given a formula F, is it unsatisfiable? What is the complexity of resolution methods that solve this problem? The resolution method used is a widely popular theorem-proving procedure; thus, the following results must be considered quite general.

Given two clauses C and D such that there exists exactly one literal q in C that appears complemented in D, the *resolvent* of C and D is defined as

$$(C - \{q\}) \cup (D - \{\neg q\}).$$

In a certain sense the resolvent is obtained annihilating the literal q. A *resolution tree* from a formula F is a tree whose leaves are the clauses of F and the son of two nodes is the resolvent of the clauses represented by the two nodes. The root is the *theorem* of the resolution tree. In figure 6 an example of resolution tree is represented.

It is easy to verify that a formula F is unsatisfiable iff there exists a resolution tree from F whose theorem is Λ. The example of the figure shows that the formula $\{\{a,b,c\},\{\neg e,\neg c\},\{e,\neg g\},\{\neg b\},\{\neg a\}\}$ is unsatisfiable. A *resolution refutation* of formula F is a resolution tree from F whose theorem is Λ. The complexity of a resolution refutation is the number of distinct clauses occurring in the tree. In the previous example the complexity of the resolution refutation is 11.

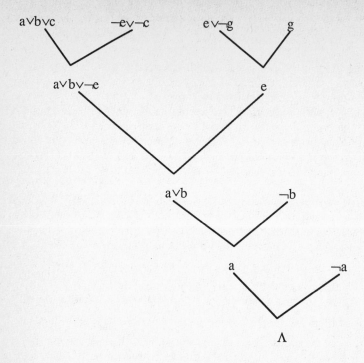

Fig. 6 An example of resolution tree: the formula is unsatisfiable

The main result of this section is that there exists an infinite sequence F_n of formulas of length $O(n)$, such that any resolution refutation of F_n has complexity Ω (c^n), for some costant c. The formulas F_n are based on graph theory. More surprisingly, they admit short refutations in axiomatic systems of propositional calculus. In particular, it can be shown that any F_n admit a resolution refutation of length $O(n^4)$ in Kleene's axiomatic system for propositional calculus.

7. CONCLUSION

In conclusion, the theory of computational complexity is a branch of theoretical computer science which has been applied successfully in the last decade to several fields. Its scope is widening more and more as new developments take place in many areas of computer science like computer system architectures, design of algorithms, or artificial intelligence and several challenging problems are facing the researchers. The goal of this introductory paper was to show how questions concerning complexity theory have been posed in several and recent areas of computer science, and how the corresponding answers have conditioned the development of research in these areas.

REFERENCES

[1] M. BLUM A machine independent theory of complexity of recursive functions, **J.ACM** **14:2**, 322-336, 1967.

[2] S.A. COOK The classification of problems which have fast parallel algorithms, **Lecture Notes in Computer Science 158**, Springer-Verlag, Berlin, 78-93, 1983.

[3] M.R. GAREY and D.S. JOHNSON **Computers and intractability: A guide to the theory of NP-completeness**, Freeman, New York, NY, 1969.

[4] L. M. GOLDSCHLAGER A universal interconnection pattern for parallel computers, **J.ACM 29:3**, 1073-1086, 1982.

[5] J. HARTMANIS and R.E STEARNS On the computational complexity of algorithms, **Trans Amer. Math. Soc. 117**, 285-306, 1965.

[6] M. LEE and Y. YESHA Separation and lower bounds for ROM and nondeterministic models of parallel computation, **Information and Computation 73**, 102-128, 1987.

[7] Y. TAO, H. ZHIJUN and Y. RUIZHAO Performance evaluation of the inference structure in expert system, **REASONING**, 945-950,1987.

[8] A.M. TURING On computable numbers, with applications to the Entscheidungs problem, **Proc. London Math. Soc. (2) 42**, 230-265, 1937.

[9] A. URQUHART Hard examples for resolution, **J.ACM 34:1**, 209-219,1987.

MEASURES OF BIOLOGICALLY MEANINGFUL COMPLEXITY

H.Atlan

Department of Medical Biophysics and Nuclear Medicine
Hadassah University Hospital, Ein Kerem, Jerusalem, Israel.

In the 1950's, John Von Neumann predicted that the elucidation of the concepts of complexity and complication would be the task of science in the 20th century as it was in the 19th century for the concepts of energy and entropy.

It appears that this prediction will come true, judging by the multiplicity of meetings and publications on these topics.

We are still at the stage where the problem is stated on the basis of difficulties and intuitions encountered in different fields such as biology, psychology, computer sciences and physics.

I shall not attempt to cover all the various aspects of what is currently emerging as the different sciences and techniques dealing with complex systems. I should like to concentrate on the differences and similarities among the problems encountered in the building of complicated artifacts on the one hand, and in the understanding of complex natural systems on the other, such as the ones we observe and analyze in biology.

The plan of my presentation is as follows:
I shall first stress the differences between the complication of artifacts and the complexity of natural systems, based on classical theories which have been proposed to measure them.

I shall then try to approach the problem of meaning, for the time being restricted to objective meaning of machines or biological systems and consider the linguistic meaning only in as much as it can be reduced to the function of a machine or organ - the language. This is obviously a poor picture of our subjective experiences of linguistic meanings and yet it is neither trivial nor devoid of interest.

I shall present this problem of meaning by way of two different approaches:
- one is derived from classical theory of algorithmic complexity and has led us to propose a measure of what we call "sophistication".
- the other makes use of computer simulations of self-organizing automata networks.

Let us begin first with the difference between the complexity of artifacts and that of natural systems.

The first has to do with our difficulties in building something, in solving a problem in order to achieve some task. The second has to do with our understanding of natural phenomena and, more specifically, of organized systems that we can observe in nature and try to analyse. It is true that at the end we hope to understand the natural systems so well that we may eventually be capable of building them ourselves. In both cases, our experience of complexity has a negative aspect: it is more that of a difficulty, a lack of simplicity.

However, in artifical systems, this difficulty is measured - to put it plainly - by decomposing the task or the problem to be solved in elementary steps that we know how to solve, and by counting the number of those steps necessary to perform the task. The situation becomes somewhat more complicated when questions of decidability arise, that is, when we do not know if we know how to do something until we try, without guarantee of success within a reasonable amount of time. From this point of view, the question of complexity of artificial systems, as dealt with in the theory of algorithm complexity, may come close to that of natural complexity.

In the case of natural systems, when trying to understand their complexity, we usually sense a gap between what we have to explain and the best explanation we can give. I once defined natural complexity as an apparent disorder where we have reason to believe that there is an order. This order is hidden from us, but, in general, we suspect, based on the observation of a function, that it exists: the apparent disorder seems to do something in an organized way, which makes some sense. As we shall see further, this lack of knowledge or apparent disorder can also be formalized under certain assumptions and be used as a measure of natural complexity.

It would seem appropriate to use different names to distinguish between the two: I suggest calling complication the number of necessary steps to solve a problem or build an artefact, and complexity the difficulty in understanding natural systems. It is not only a question of semantics, since the problems and the necessary approaches are different, as we shall see, although in the end they may come together under certain conditions.

The complication is dealt with in the theories of algorithm complexity, which have been developed essentially to meet the needs of computer sciences and programming. Schematically, the complexity of an algorithm aimed at performing a given task is measured by the time it takes to perform it, measured in units of computer instructions. Since this time is going to depend upon the computer's capabilities, one uses a kind of theoretical normalized universal computer, the Turing machine, which is capable of doing in principle, by very elementary steps, everything a digital computer can do. What is important is not so much the number of program instructions or the computation time in absolute value, but rather how this time varies as a function of the size of the problem to be solved. The problem size is defined by a number n, which is the number of input data to be processed in the problem (for example, number of variables and parameters, number of elements in a matrix, number of nodes in a network, etc.). The computation time is obviously a function of the problem size, and the question is, for a given algorithm, which kind of function it is. The more this function increases rapidly with the problem size n, the more the algorithm is considered to be complex. The simplest are those which vary like the logarithm of n, or like n itself; more complex are those which vary like a power of n or polynomials of increasing degrees in n. The most complex are those which increase like exponential functions of n, and for all practical purposes are often considered insoluble.

Many efforts have been made to know, for a given problem, whether or not it is possible to find an algorithm capable of solving it as well as what would be a minimum value for its complexity. It is not my intention to summarize in any way what is being done in this field. I am sure we shall hear more about it during the course of this meeting. My only purpose in mentioning it is to suggest that the problems involved in this discipline are very different from those encountered when we face the complexity of natural systems, even though, as we shall see later, something can be learned from it, by difference more than by analogy, especially when we are using computer simulations of natural systems.

A simple way for us to realize now that the complication of a given task does not account for the natural complexity which results from our difficulties in understanding will be to quote the conclusion of a chapter on algorithm complexity in a textbook (0) on The Design and Analysis of Computer Algorithms: "It follows from Theorems 1.1 and 1.2 that as far as time complexity (and also space complexity is concerned, the RAM and RASP models are equivalent within a constant factor, that is, their order-of-magnitude complexities are the same for the same algorithm. Of the two models, in this text we shall mainly use the RAM model, since it is somewhat simpler" (emphasis added).

RAM and RASP are the names of two computation models or elementary languages that are equivalent - or, more precisely, polynomially related - to Turing machines. That is to say, every algorithm, the complexity of which is a polynomial function of n in one of these languages, will also have a complexity of the same order of magnitude in the other and in a program implemented on a Turing machine. These two models are different in that the first one, RAM (random access machine or memory), does not store the program instructions in memory, so that they cannot be modified by further instructions in the program itself. However, it can make use of indirect addressing. On the contrary, the second model, RASP (random access stored program machine), does not need indirect addressing because its program instructions are stored and can be modified.

What these theorems are saying is that these two computation models or languages are equivalent from the point of view of the complexity of the algorithms that one can write with them. However, the last sentence in the text tells us that, in a sense which is necessarily different from what we just said, one of them, the RAM is "somewhat" simpler, that is, less complex than the other. In fact, this "somewhat" refers to the fact that the textbook is now using a metalanguage to help students understand the theory, and it is in this metalanguage and from the point of view of ease of explanation and understanding that the RAM seems to be, in an intuitive, not defined way, less complex than the RASP. Thus, it is obvious that this complexity of understanding is different from the one previously defined in the mathematical theory of algorithm complexity, the one I suggested be called complication.

Attempts have been made to measure the complexity that appears in the analysis and understanding of natural systems, in particular, living ones, - whether single cells or organs like the central nervous system or the immune system, organisms, or ecological systems. Most often, a probability function called entropy, information content, or uncertainty, taken from Shannon's information theory, is used as a measure of complexity. It is important to bear in mind that this Shannon function, even though counted positively, expresses in fact a negative quantity: it is an a priori uncertainty, a lack of knowledge, or missing information about how to build a system from its elementary parts, similar to the one we observe in nature. And it is based on the probability that such a system may have been assembled randomly. The less probable that is, the more we need information to reproduce it. But we obviously do not assume that a given organized and functioning system has been assembled at random. Natural systems are assumed to be the products of physico-chemical determinations, whether complete or incomplete, at work in nature. Only, we are missing part of the knowledge and the understanding as to how they occurred and how we should proceed if we should want to reproduce them ourselves. Whether this lack of information is accidental - the determinisms are supposed to exist but we don't know them in detail or essential, that is, some intrinsic indeterminations would exist in nature at the level of observation under consideration, is a metaphysical question that is not relevant in this context.

The point is that this uncertainty function of Shannon measures the missing information, whatever its origin. It measures our missing information as to the laws and determinations that would allow deduction with certainty of the structure and the function of a system when the properties of its parts are known.

It is possible to define a quantity called redundancy which is the opposite of complexity, and represents contrary to missing information and uncertainty, known constraints and observed regularities.

TABLE

Redundancy	Complexity
Repetition	Variety
Symmetry	Asymmetry
Homogeneity	Heterogeneity
Degeneracy	
Unspecific	Specificity
Undifferentiated	Differentiation
Sameness	Difference
Interchangeable	Individuality
Repetition in time	Unexpectedness
(⇒ causality,	(⇒ newness in time,
determinism)	uncertainty, randomness)

I have listed here on the table how different notions which seem close to one another distribute themselves in antagonist couples and are opposed to one another just like complexity and redundancy.

In spite of many well-recognized insufficiencies, one advantage in using this probabilistic approach has been to allow for a formalization of the apparently paradoxical role of randomness or noise in self-organization processes.

Self-organization appears to be a property of several natural systems, such as the central nervous system and the immune system in mammals, where not everything is genetically determined and where encounters with non-predicted events play an important role in their epigenesis; or of ecosystems, where self-organizing properties are even more obvious since, in that case, no genetic determinations can exist at the level of the whole system; and also of any cognitive system human and non-human - capable of non-directed learning: whereby the organization of the system assimilates its environment (in the two senses of physiological and cognitive assimilation, as in Piaget) in such a way that it is the result - that is, the organization itself is the result - of the learning process.

It was possible to use and extend probabilistic information theory to lay out a formal mechanism of self-organization whereby noise in the form of random perturbations is utilized by a sytem up to a certain point to increase its complexity at the expense of its redundancy, and increase its capabilities of adaptation to new unexpected environmental factors.

I will not recall the details of this work known as a kind of complexity from noise principle, which has been published extensively (1-4).

My purpose today is to concentrate on a problem or a difficulty which is common to these two measures of complexity, of algorithmic artefacts and of natural systems, in spite of their obvious differences.

Both measures fail to take into account the meaning of what is measured. In the case of a computer program, it is obvious that it always has a given meaning, namely, solving a given problem or achieving a given task. Nevertheless, the theory of algorithm complexity is based on logical considerations of program length and decidability which do not have to explicitly take into account - let alone measure - the actual meaning of a program in terms of setting up what it is good for, i.e. stating the problem to be solved or the given task to be achieved.

As a conspicuous consequence according to the commonly accepted definition of algorithm complexity or complication by the program length necessary to perform a task, a given infinite random string is the most complicated task to perform since there is no way of generating it by means of a program shorter than the actual string itself. The same problem is found in the measure of a natural object complexity by Shannon's entropy: as any entropy it is a measure of disorder in the sense of random homogeneity.

The only difference between a natural complex system and plain disorder and randomness is our experience - or our assumption, more or less justified - that the former performs a function, i.e. is endowed with a certain meaning. This is particularly clear in biological applications of concepts borrowed from information theory: one of their basic flaws stems from the fact that biological information has a meaning, which manifests itself in physiological functions. For example, the meaning of genetic information is to be looked for in the phenotype, i.e. in the phenotypic characters of the organism both morphologic and functional. Thus, answering the question of how the meaning of biological information is generated amounts to understanding how physiological functions observed at an integrated organismic level are generated from information stored and processed at a molecular level.

At the beginning of molecular biology this question seemed to be straightforward, since it was answered by the known dogma of molecular biology: 1 gene → 1 enzyme → 1 character: the meaning was supposed to be a one-to-one application of gene structures into phenotypic characters. We know today that it is not so easy because in most cases many genes contribute to one character and a single gene participates in the determination of many characters.

This is where the approach of this problem by simulations of automata networks is useful because it helps in understanding generic properties of large numbers of interacting genes, as they appear at the macroscopic integrated level of the behavior of the whole network.

Also, the idea of complexity from noise and the use of Shannon theory to describe processes of self-organization with increase in complexity relies on the assumption that new meanings are generated, although we do not know how.

Otherwise, the idea that noise may help to generate new organization would amount to saying that, for example, by shaking and perturbating at random a refrigerator, one may end up with a new machine, for instance a dishwasher! The idea of self-organization as a process of constant disorganization accompanied by reorganization, necessarily amounts to constant creations of new meanings which allow us to talk about new organizations.

Until recently, the fact that both classical measures of algorithmic complexity and of entropy have not dealt explicitly with what makes them meaningful, was not considered to be a basic flaw. Because the meaning is always assumed to exist, it is implicit and one just disregards the case of plain randomness.

One knows very well that a program is not a random string, because it does something, although formally, if it is a minimum program, it is random string in the technical sense that it is not compressible.

This is why one does not bother with this idea that technically maximum randomness amounts to maximum complexity.

Now, it we wish to go further, we must try to formalize the generation of meaning which has to do with <u>performances of functions</u>. I am referring not only to physiological functions of organisms but also to performances of machines and algorithms where the meaning of a part or of a program instruction has to do with its function, i.e. with what it is good for within the overall functional organization - the flow chart if you wish - of the system.

However, it is important to realize that the meaning of information in natural organizations such as biological systems is very different from what it is in artificial constructs: in man-made machines or programs the goal is defined from the beginning and it is in relation to this goal that parts, connexions, signals, etc. either have a meaning or are meaningless. In a natural organization, the final goal, if there is one, is not known, and meanings are defined a posteriori according to <u>self generated</u> criteria. This is why a real understanding of self organization would amount to understanding mechanisms by which <u>non-purposeful systems</u>, not goal-oriented from the outside, organize themselves in such a way that the meaning of information is an <u>emerging</u> property of the dynamics of the system.

Before attempting to show you how computer simulations of automata networks can help to visualize such mechanisms, I should like to spend some time on a recent attempt in collaboratoration with M. Koppel, to formalize meaningful complexity keeping within the framework of the theory of algorithms (5,6).

The approach is based on the following question: faced with a natural phenomenon which seems complex to us - because it does something or for some other reason - is there a way to decide to what extent we are dealing with randomness or with a complex meaningful structure? In other words, in what we observe, what part is played by randomness and what part by meaningful structure?

This question is not only a formal one: for example if we receive signals from outer space or more practically when we look at DNA sequences, is there a way of knowing how much meaningful complexity that is worth investigating there is?

What is the relationship between intrinsic properties of a string e.g. the outer space signals or the DNA sequence - and its function, i.e. its effect on a receiver, that usually tells us about its meaning?

In our preliminary attempts to answer this question we based ourselves on a modified version of the classical Kolmogorov-Chaitin theory of algorithmic complexity.

According to this theory, the complexity of a string - such as the ones I just mentioned - is the length of the minimum description capable of reproducing it when it is given to a normalized general computer, i.e. a Turing machine. This description is itself a string which can be eventually written in a binary alphabet and the number of bits is a convenient measure of its length.

Now, in this description, no distinction is made between what is a program and what are data.

When we are dealing with usual man-made procedures, the distinction is obvious: a set of operations is a program and the data on which it operates are the data.

However, for the theory, it does not make any difference, because for the machine itself it does not make any difference: the bits of a program are themselves data for the interpreter of the machine and therefore it seems that there is no intrinsic difference between computer programs and data. In fact, the realization that there is no intrinsic difference for the machine betwen program and data was a great achievement which has led to highly significant breakthroughs in the theory of computation.

At the same time, however, something essential has been lost. Because this is true, again, only if we agree to put aside the distinction between what is meaningful and what is not in the minimal description of a string. If we wish to capture the meaningful structure of the string, we must come back to the distinction between program and data, and make it even more clear cut. What is at the root of this distinction is the fact that a program has a certain range, i.e. the class of all the obj ects or strings it can produce when it is given all possible data.

Thus what is common to this class of objects is their structure or meaningful complexity, whereas each particular element of the class is a particular instance of this structure.

Then, instead of considering a usual Turing machine, we consider one with two distinct input tapes- one for the program and one for the data. And the minimal description of a string is thus made of two parts, program and data, each with its length, so that the usual complexity measure H (S) becomes the minimum of (length of program P + length of data D) (P,D) being a description of the object S:

H(S)
= min {|P| + |D|(P,D) generates S when input to a universal Turing machine}
length of program and length of data where (P,D) is a description of the object

Now, what we call the meaningful complexity of S, or as we called it, its sophistication, is only the length of P, i.e. the length of the program part in the minimal description. Thus, we succeed in sorting out the meaningful complexity in such a way that a random string where the program part is zero has a zero sophistication, while its classical complexity is high.

In fact, it is not that simple, because the fact that data and program together can be regarded as one single input i.e. data to a universal Turing machine, has deep justifications from the computation point of view, which have to do with the role of the Turing machine as an interpreter.

Therefore, it may be that a minimal description is minimal only for a given particular string and not necessarily for the whole class of strings which share the same structure: in other words, it is possible that this particular minimal description is ad hoc; and if we want one which is not, and really captures the structure of the class, we may want to allow the description to be longer than the minimum in order to incorporate a program which would be longer but appropriate for the whole class.

This is why we introduced the idea of c-minimal descriptions, made of (P,D) such that

$$|P| + |D| \leq H(S) + c$$

i.e. we allow for the description to be longer than the minimal by a constant c.

And the sophistication or meaningful complexity is defined in relation to this constant as c-Sophistication

$$SOPH_c(S) = \min\{|P| \mid \exists D, \ |P| + |D| \leq H(S) + c$$

i.e. the minimum program length P such that (P,D) is a c-minimal description of S.

The role of constant c is better understood by an example. The following are three possible different descriptions able to generate the characteristic string of prime numbers:

1. (PRINT DATA) = Prints 1,2,3,5,7,11,13,17
   (Needs data on primeness of all numbers)

2. $(P_2, \ ODD \ DATA)$ = First tests for evenness to eliminate even numbers (Greater than 2).
   Then prints the primes from the data on odd numbers.
   Needs data only on primeness of odd numbers)

3. ("PRIME") = Applies on every number a test for primeness.
   (Does not need any data)

Let us try to find the sophistication of that string by looking at what its minimal description is. This obviously will depend on the length of the string we are interested in. For a short string of a few prime numbers, PRINT + the DATA is certainly the shortest description shorter than the program PRIME. However, the longer the string, the longer the data necesssary for that description, the more economical the actual program PRIME appears, because it requires less data, as a matter of fact, none at all.

Intermediate between them is P_2: the additional program length to test for evenness is economical when the length n of the string becomes longer, so that the n/2 bits of data which are saved by the test compensated for this additional length.

Therefore, we do not always want the true minimal description such as PRINT + FEW DATA which might be ad hoc and appropriate only for a given set of data corresponding to only a few members in the class of strings which share the same structure. We allow the description to be c-minimal i.e. longer than the minimal by a constant c. This allows for additional length of program which will be confirmed when additional data are given. In that sense, c is a measure of confirmation or redundancy. If a program is the true one, it will remain the shortest even for large values of c, i.e. for long strings.

Thus, the constant c measures what we are ready to accept as additional bits in order to be confident that we are not dealing with an ad hoc minimal program, i.e. a program which works only for a given set of data.

One of the advantages of this definition is that it can be extended to infinite strings; and then it can be shown that the parameter c is eliminated.

Thus, a finite sophistication can be defined for an infinite string. It can also be shown that its value is invariant upon the choice of the Universal Turing Machine.

In cases where the length of the minimal <u>program</u> increases and tends to infinity, when the length of S tends to infinity we are dealing with an infinite sophistication and this defines in a precise way an object which is <u>neither recursive nor random</u>. It is worth noting that this strange kind of object might be the kind of thing nature seems to manufacture normally.

These definitions have been of some help in analyzing the question of what is the computing nature of the DNA strings: are they program or data? According to the widespread metaphor of the genetic program DNA are viewed as programs computed by the cell. When the metaphor is taken too seriously, it leads to the false assumption that knowing the DNA sequence of a species is enough to describe its structure and function. It seems more likely that DNA strings should be looked at as data fed to the cell metabolism machinery which would work like a program. If that is the case it is obvious that the assumption which underlies the much publicized project of Sequencing Human Genome is wrong.

For the last part of my talk, I would like to go back to the emergence of meaning in functional self-organization to show you that it is not so mysterious, since it can be simulated relatively easily by networks of automata. This work This work was done in collaboration with Weisbuch and Fogelman (7-10), and made use of boolean networks which were first studied by Kauffman (11).

Networks are constructed, made of connected boolean automata; that is, each automaton is located at a node of a 16 x 16 matrix closed on a torus; it receives two input connections from two of its opposite neighbors and sends two output connections to its two other neighbors. Each automaton is characterized by a given boolean function of two variables.

At every step each automaton updates its state by computing its function from its two binary inputs and its new state is sent as output. The 14 non-constant two-variable boolean functions are distributed randomly on the nodes of a matrix. Initial states are also set randomly and the automata compute their new states in parallel, at every time unit.

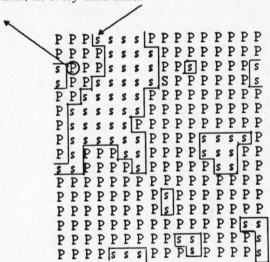

Fig. 1: Example of network in its limit cycle exhibiting a macroscopic spatiotemporal structure. The two arrows indicate input and output elements of a recognition channel. Perturbating binary strings are imposed on the input element and stabilization of the output element by a given string is interpreted as "recognition" of the string.

Well-established results have shown that in most cases the network reaches a limit cycle relatively rapidly. In its limit cycle the network presents a macroscopic spatiotemporal structure since it is divided into connected subnets, stable and oscillating. An example is shown in fig. 1 where S represents the stable elements and P periodically oscillating elements. This process is an instance of structural self-organization with emergence of macroscopic structures from local interactions, which has been extensively studied under different conditions as far as the nature of the automata and of their connections are concerned. Less known is the utilization of such networks for simulations of functional self-organization. When such a net is in its limit cycle, temporal binary sequences are imposed on one automaton and work as "permanent" perturbations of the limit cycle. By "permanent" we mean a duration of two cycle lengths, plus an aditional two cycle lengths, during which the perturbed state of the network is observed. Because of its robustness the macroscopic structure of the limit cycle is very often grossly conserved (depending on the location of the perturbed automaton), with the exception of a few stable elements which are destabilized and also, surprisingly enough, some oscillating elements which are stabilized by the perturbations. Since stabilization of a given oscillating element by imposition of perturbating sequences on another one occurs only for a given class of such sequences, this process can be used as a sequence recognizer. The interesting feature in this phenomenon is the definition of the criterion for the classification of sequences into those which are recognized (by the fact that they stabilize a given element) and those which are not. An example of such criterion is shown in figure 2. Classes of binary sequences are defined by partially random periodical structures where indifferent digits (as those indicated by stars in fig.2) are inserted at given locations.

```
a    00101000100000011010000110001001
b    10001001000010010010100100101000
c    01001100100010001011010001000101
d    10100001101010010010100110101000
```

```
     *0*0*00*
```

Fig. 2: Example of criterion for recognition characteristic of a given pair of input and output elements of a structured network in its limit cycle. This criterion, as the macroscopic structure of the limit cycle itself, is an emerging property of the dynamics defined by the microstructure of the network (laws and connections) and its initial state.

Strings a, b, and d were able to stabilize the output element (they were "recognized") when imposed on the input one of a given channel in a given limit cycle, because they exhibit the partially random periodical structure of 8 digits represented below. The star elements are indifferent and may be randomly set. One can verify that string c which was not recognized does not match this criterion.

The definition of such a structure is the result of a given pathway in the network between the input element where the perturbating sequence is imposed and the output element which is (or is not) stabilized by the sequence. The existence of "star" indifferent elements is the result of the overall spatiotemporal structure of the net in its limit cycle and of the "reducing" effects of computation by some of the boolean functions which associate one identical output to three different two-variable inputs, as is shown in fig. 3.

Thus, a specific pathway in the network after it has reached its final structured state, between a given element serving as input and another one serving as output, defines a class of sequences to be recognized.

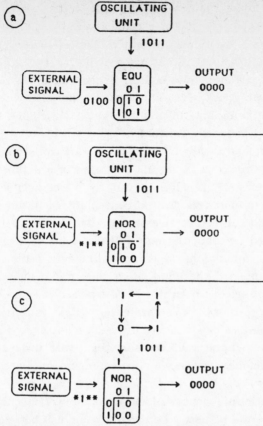

Fig. 3: from [7]. Mechanisms of stabilizations depending on the kind of boolean functions computed by an element which receives one input from outside (the perturbating string) and the other from the rest of the network in its limit cycle.

a) An oscillasting EQU element can be stabilized into a constant 0 state by a single perturbating sequence applied to one of its inputs which matches exactly the sequence of the other input. b) The NOR element recognizes a class of partially random sequences. The first, third and fourth bits of sequences can have any value since the element is then stabilized in 0 state by the input 1 of an oscillating unit produced by the network in its limit cycle. This property is shared by all three-to-one reducing functions.

c) Some two-to-one reducing functions depend on one input value only. They merely transfer this value (or its dual) from one element to the other without being influenced by the second input. Their role is to transfer sequences from one place in the network to another, thus sometimes making the effects of a perturbating string felt far away from the input element. In addition, they create loops which, if frustrated, work as oscillating units which generate the oscillating behavior of subnets in the limit cycle. A recognition device can be designed from a generating frustrated loop made of such transfer functions associated with a 3-to-1 reducing function for fuzzy recognition (i.e. classes of partially random strings), or a non-forcible function (EQU, XOR) for precise recognition (limited to single tryly periodic strings).

In other words, the structural self-organizing process results in a non-programmed definition of criteria for classification of binary sequences (9,10).

This model shows a mechanism by which a set of messages is divided into those that are recognized and those that are not, while the criterion for this demarcation - which is similar to making sense and not making sense to a cognitive system - is nothing other than a given inner structure, which has no other meaning than being able to produce this demarcation, and, itself may have come about, at least partly, randomly.

It is as if complexity, which appears as an apparent non-reducible randomness, can be removed by means of a kind of orderliness that did not come about as a result of planning but as a result itself of apparent indeterminacy and randomness. This, in my opinion, is the consequence of the close relationship between complexity and disorder in natural systems not planned and ordered by man, the only difference being the existence of an apparent meaning or function in the former in the eyes of the observer.

This may also explain the feeling that we have about how nature seems to build its machines. This has been described by Francois Jacob as "bricolage" ("thinkering"). To perform a complicated task, such as embryonic development or evolution of species, nature does not seem to seems to take what happens to be at hand and does something with it depending upon the circumstances, sometimes in a very cumbersome and not necessarily the most economical, manner.

We get the same feeling when we look at our networks and see the kind of criteria which happen to be used in order to distinguish between classes. They also seem to be farfetched from our point of view, if we would have to set up such criteria for what makes sense and what is meaningless.

REFERENCES

0. A.V. Aho, J.E. Hopcroft & J.D. Ullman. The Design and Analysis of Computer Algorithms. Addison Wesley, Reading, Mass., 1974, p. 19.
1. H. Atlan. On a Formal Definition of Organization. J. Theoret. Biol. 1 974, 45, pp. 295-304.
2. H. Atlan. L'organisation biologique et la theorie de l'information, Paris, Hermann, 1972.
3. H. Atlan. Entre le Cristal et la Fumee, Paris, Seuil, 1979.
4. H. Atlan. Hierarchical Self Organization in Living Systems. Noise and Meaning, in: Autopoiesis: A Theory of Living Organization, ed. M. Zeleny, N.Y., North Holland, 1981, pp. 185-208.

5. M. Koppel, Structure, in: <u>The Universal Turing Machine, A Half Century Survey</u>, R. Herken ed., Oxford Univ. Press, London, 1987.

6. M. Koppel and H. Atlan. Program Length Complexity Sophistication and Induction, in preparation.

7. H. Atlan, F. Fogelman-Soulie, J. Salomon and G. Weisbuch, Random Boolean Networks, <u>Cybernetics and Systems</u> 12, 1981, pp. 103-121.

8. F. Fogelman-Soulie, Frustration and Stability in Random Boolean Networks, <u>Discrete Applied Mathematics</u>, 9, 1984, pp. 139-156.

9. H. Atlan, E. Ben Ezra, F. Fogelman-Soulie, D. Pellegrin and G. Weisbuch. Emergence of Classification Procedures in Automata Networks as a Model for Functional Self Organization, <u>J. Theoret. Biol.</u>, 1986, 120, pp. 371-380.

10. H. Atlan. Self Creation of Meaning, <u>Physica Scripta</u>, 1987, 36, pp. 563-576.

11. S. Kauffman. Metabolic Stability and Epigenesis in Randomly Constructed Genetic Nets, <u>J. Theoret. Biol.</u>, 1969, 22, PP. 437-467.

Complex systems, organization and networks of automata.

Gérard WEISBUCH, Groupe de physique des solides de l'Ecole Normale Supérieure, 24 rue Lhomond, F-75231 Paris Cedex 5, FRANCE.

Although some people even question the interest of a concept of complexity, I definitely use it and I consider a system to be complex if it is composed of a large number of different elements in interaction. The three qualifiers are necessary, namely: large number of elements, different elements, and interactions. Of course I have in mind biological systems but my approach is a generalisation of that of the physicists of disordered systems in the sense that I am mainly interested in dynamical properties. Among the possible simplifications necessary to deal with complex systems, one consists in maintaining a large number of otherwise simplified elements. Discretization is the heart of the theory of networks of automata.

Definitions

Structures

An *automaton* is a binary device which computes at *finite time steps* its own binary state as a function of binary *input* signals coming from interacting automata. This function is called the *transition function*. We shall only consider the case where the state of the automaton is also its *output*. This definition is a simplification of the definition used in computer science.

A *network* (or a net) is obtained by the connection of several automata. The inputs to update the state of one automaton at time t are the states (or outputs) of the connected automata at time t-1. According to the application, automata might model for instance genes, nervous cells or chemical species, while the network models a living cell, the central nervous system or a primeval soup at the origin of life.

The *connection graph* is the set of connections established between the automata. This graph might be complete (all automata connected), random or regular. The latter case is that of cellular automata, more frequently used for technological applications.

Different types of automata might be considered according to their transition function.

The general case is that of *Boolean* automata. They operate on binary variables, whose values are 0 or 1. Usual functions in logics, AND, OR, XOR, are examples of transition functions depending upon 2 inputs. A Boolean automata with k inputs is defined by a *truth table* which gives the state of the automaton as a function of the 2^k input configurations. The transition function of a boolean automaton with k inputs can be any of the 2 to the 2^k Boolean functions with k inputs.

Finally, for a given network, several modes of operation can be used to apply the transition rules: either all automata apply simultaneously the transition rule, which is called *parallel* iteration, or this process is done sequentially, one automaton at a time, in a *sequential* iteration mode.

Dynamical properties

Once a network is defined by its connections among the automata, by the transition rule for each automaton and by the choice of an iteration process, one is interested in its dynamical properties, i.e. the description of the configurations reached by the network when the transition rules of the automata are applied at regular time intervals. For networks made of a small number of automata, say 10, the complete *iteration graph* (the set of all the configurations of the net with arrows pointing to the successor of each state) can be drawn. Several concepts can be defined:

Attractors. If starting from an initial configuration, a configuration is reached twice, the net indefinitely cycles through the subset of configurations between these two occurences. This subset is an attractor of the dynamics. Attractors composed of only one state are called *limit points*, the others being called *limit cycles*. The number of states of an attractor is its *period*. It is the time separating the occurence of the same configuration.

Transients. The states not belonging to any attractor are called transient. The subset of all transients which evolve towards the same attractor plus the attractor, is called an *attraction basin*.

These definitions are of course valid for nets of any size. Since the number of configurations of a net varies exponentially with its size (2^N for size N), it is impossible to know the iteration graph for large N.

In fact for large nets several possible behaviors can be imagined:

Organized behavior : the large number of possible is strongly reduced after several iteration steps to only a small fraction of the configuration space.There exists a small number of attractors with a short period which scales as a power of N, the number of automata of the net. Formal neural nets as described by D. Amit in this volume give an example of such a behavior: their number of attractors scales as N, and their period is 1.

Chaotic behavior: on the opposite, one can imagine the case when a large fraction of the configuration space is describes between the re-occurence of the same configuration. The periods scale as an exponential function of N. By analogy with the continuous models this regime is called chaotic.

Both regimes were reported by S. Kauffman as early as 1969 [1].

Random Boolean Nets

In a few cases our knowledge from biology is sufficient to build a model with an explicit interpretation for every automaton and its interactions, and for the behavior of the system. But most often this is not the case, especially for large complex systems and one is interested in their generic properties. Finite discrete systems, such as networks of automata are in a finite number. The probability of occurence of some chosen behavior can then be defined. The generic properties of the systems are those qualitative properties which occur with a probability infinitely close to one. Or, in the case of semi-quantitative properties like scaling laws, one considers the average dynamical quantities. Of course one does not need to conduct an exhaustive study of all nets built from some set of rules, but it is suficient to work on randomly generated nets.

A very general approach has been proposed by Stuart Kauffman [1]. It consists in considering the dynamical properties of random nets, composed of Boolean automata with transition functions randomly chosen among Boolean functions with k inputs (k, the connectivity being constant), and with random connections. The parallel iteration mode is selected. The question is to determine whether there exist properties which are generic on the set of random nets: generic in the sense that they are exhibited by almost all nets, possibly with few exceptions corresponding to special designs.

Among the properties that were first exhibited by computer simulations are the following:

There exists a transition in behaviour between nets with connectivity 1 or 2 and those with larger connectivity. When one increases the number of automata, for low connectivities small periods and small numbers of attractors are observed, whereas exponential growth of periods is observed in the opposite case which is of no interest in biological modeling.

For k=2, the period and the number of attractors vary as the square root of the number of automata. During the limit cycles some automata remain stable while others are oscillating. The set of stable automata is often strongly connected and isolates subnets of oscillating automata.

Kauffman proposed to model cell differentiation in ontogeny from a common genome by random Boolean nets. The expression of the genes inside a given genome depends upon the concentration of regulatory proteins synthetized by other genes. The fact that a gene is expressed or not can be represented by a binary state, the interactions among genes via the proteins by Boolean automata with their connection structure, and the genome by the Boolean network. Within such a formalism, the different cell types are interpreted as the different attractors of the net. This interpretation is supported by the fact that the cell division time and the number of different cell types scale as the square root of the DNA mass of the genome, the same law as for the corresponding quantities for Boolean nets with connectivity 2.

Kauffman's simulation results were never exactly derived in spite of numerous theoretical attempts. In this talk I shall discuss various approaches to compare the dynamical behaviors in the two regimes, including the various scaling laws, and I shall attempt to describe the nature of the transition among them.

Spatial organization

The existence of short periods in the organized regime can be interpreted in terms of temporal organization. A correlated functionnal organization can be observed with cellular lattices.The first cellular implementation of Kauffman [1] nets on a cellular lattice is due to Atlan et al. [2]. It consists in placing Boolean automata with connectivity 2 at the nodes of a square lattice with a connectivity matrix described in figure 1. Since connectivity is 2, these random nets only exhibit an organized behavior.

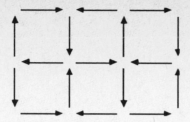

Figure 1. Pattern of connectivity for 2 inputs cellular automata on a square lattice.

One then observes that during the limit cycles some automata remain stable while others are oscillating (cf. figure 2). The set of stable automata is often connected and isolates subnets of oscillating automata. The details of the *patterns* depend upon the initial conditions and are specific of the attractor.

```
. 0 0 0 . . * . 0 1 * . 1 0 1 *        . 1 1 1 . . * . 0 1 * . 1 0 1 *
* 0 1 0 1 0 . . 0 1 * . 1 1 1 1        * 0 1 0 1 0 . . 1 1 * . 1 0 1 0
0 . . 0 0 0 0 . . 0 1 1 * 0 1 1        0 . . 1 1 0 1 . . 0 1 1 * 0 1 1
. . 0 0 1 0 0 * * 1 1 0 * . . * .        . . 0 0 1 0 0 * * 0 1 1 * . * .
* * 0 1 * . 0 0 . 0 * * . * * .        * * 1 0 * . 0 0 . 0 * * . * * .
. 0 0 0 . . 0 1 . * 1 0 * . . *        . 0 0 0 . . 0 0 . * 1 1 * . . *
. . * 0 . * 0 0 1 . 1 1 1 . . .        . . * 0 . * 0 0 1 . 1 1 1 . . .
. 1 0 0 . 1 0 . * 0 0 * 1 . . .        . 0 0 1 . 0 0 . * 1 0 * * . . .
0 1 1 1 0 1 1 0 0 * * * 1 0 * 1        . 1 0 1 1 1 1 0 0 * * * * * * .
1 * 0 . 0 1 1 0 1 1 . 1 1 1 0 0        * * . . 0 1 1 1 1 0 . * . * * .
0 0 1 * 1 0 0 1 * . 1 0 . 1 1 1        . * * * 1 0 0 1 * . 1 1 . . * .
1 1 * * 0 1 0 . * * 0 0 * 1 1 0        . * * * 0 1 0 . * * 0 1 * * . .
* . * * 0 0 0 * . . . 0 0 1 0 .        * . * * 0 0 0 * . . . 1 0 1 1 .
* * * . * . . . . * . 1 0 . *        * * * . * . . . . * . . 1 0 . *
. * . . . * . * * * . 1 0 1 . .        . * . . . * . * * * . 0 1 1 . .
* . . * * * * * * . . 1 1 . . .        * . . * * * * * * . . 1 1 . . .

. 0 1 0 . . * . 1 0 * . 1 1 1 *        . . * . . . * . 1 0 * . 0 1 0 *
* 0 0 0 0 0 . . 1 1 * . 0 1 0 1        * . * . * . . . 0 0 * . 0 1 1 1
0 . . 0 1 0 1 0 . 1 1 1 1 1 1 0        1 . . . * . * 1 . 0 1 1 1 1 0 0
. . 1 0 0 0 1 1 0 0 1 1 0 * .        . . . . . * . . 1 0 1 0 0 0 0 * .
* * 1 1 * . 0 0 0 1 1 0 0 * * .        * * * * * . 0 0 0 0 1 0 0 * * .
. 0 1 0 . . 1 0 0 1 0 1 1 . . *        . . . . . . 1 1 1 1 1 0 0 . . *
. . * 0 . * 0 0 1 . 1 0 1 . . .        . . * . . * 0 1 1 . 1 0 1 . . .
. 1 0 0 . 0 1 . 1 1 1 * 1 . . .        . 0 1 1 . 1 1 . 0 0 0 * 1 . . .
0 0 1 0 0 1 1 1 0 * * * 0 1 * 0        0 0 1 0 0 0 1 1 0 * * * 0 1 * 0
1 * 0 . 1 1 0 1 0 0 . 1 0 1 1 0        1 * 0 . 1 0 0 0 1 1 . 1 0 1 1 0
1 1 0 * 1 1 0 0 * . 1 1 . 0 0 0        1 1 0 * 1 1 0 0 * . 1 1 . 0 0 0
0 1 * * 0 0 1 . * * 0 1 * 1 0 0        0 1 * * 1 1 0 . * * 0 1 * 1 0 0
* . * * 0 1 0 * . . . 1 0 1 0 .        * . * * 0 1 0 * . . . 1 0 1 0 .
* * * . * . . . . * . . 0 0 . *        * * * . * . . . . * . . 1 1 . *
. * . . . * . * * * . 0 0 1 . .        . * . . . * . * * * . 0 1 1 . .
* . . * * * * * . . 0 0 . . .        * . . * * * * * * . . 1 1 . . .
```

Figure 2. Patterns of activity of the same 16*16 net during the limits cycles reached from 4 different initial conditions: The 0's and 1's correspond to oscillating automata, while . and * correspond to automata that remain fixed.

A possible analysis of this dependence on initial conditions, is to summarize, as in figure 3, how many times each automaton is oscillating for a given number of initial conditions (see Atlan et al. [3]). Figure 3 shows that a large proportion of automata remain always stable, the *stable core* [4].

31	533	533	533	0	0	0	0	999	999	0	0	999	999	999	31
31	533	533	533	533	533	0	0	999	999	0	0	905	999	999	999
999	0	0	533	533	564	564	749	0	999	936	936	717	936	999	999
0	0	564	564	533	564	564	749	749	999	936	936	655	655	0	0
0	0	564	564	0	0	999	999	749	999	718	655	655	0	0	0
0	564	564	564	0	0	999	999	749	749	936	936	655	0	0	0
0	0	0	564	0	0	999	999	999	62	936	936	936	0	0	0
0	999	999	999	0	999	999	62	780	936	936	62	407	0	0	0
438	999	999	999	999	999	999	999	62	62	62	438	438	0	438	
438	0	438	0	999	999	999	999	999	999	62	438	438	438	438	438
438	438	438	0	999	999	999	999	63	63	782	782	0	438	438	438
438	438	0	0	999	999	999	0	63	63	782	782	0	438	438	438
31	155	155	0	999	999	999	0	63	63	0	782	782	937	937	31
31	155	155	155	0	0	0	0	63	63	0	0	999	999	0	31
31	0	0	0	0	0	0	0	0	0	0	875	999	999	31	31
0	0	0	0	0	0	0	0	0	0	0	875	875	31	31	31

Figure 3 Statistics of those initial conditions (out of 999) which lead each automaton to oscillate during the limit cycle.

Those which oscillate are grouped in clusters, with contiguous but discontinuous probability. Each probability step corresponds to a different limit cycle and the intervalls between the nearest probabilities to the width of the attraction basin.

Four input square lattices.

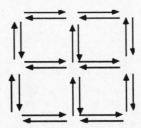

Figure 4. Pattern of connectivity for 4 inputs cellular automata on a square lattice.

A nice way to study the frozen/chaotic transition is to use a continuous parameter. Derrida and Stauffer [5] have proposed to work with 4 input square lattices. If the transition function of the automata are chosen symmetrically with respect to 0 and 1 behavior of the net is chaotic. But on can choose to bias the probability so that the transition function yields 1 for any input configuration. If this probability is 0 for instance, the net has only one stable attractor: in one iteration step any initial configuration evolves towards 0. When p varies from 0 to .5 the transition should occur somewhere (The region from .5 to 1 is symmetrical). Computer simulations in Weisbuch and Stauffer[6] show that for low values of p the periods vary slowly with N, which indicates a frozen regime, while they grow exponentially with N at larger p (chaotic regime).

The **local periods** are also quantities of interest (Weisbuch and Stauffer [6]). The state of each single automaton evolves periodically with its own local period which divides the period of the whole net. Figures 6 and 7 show these periods for both regimes. In the frozen regime oscillating automata are grouped into small clusters with medium periods. On the opposite, in the chaotic regime, automata seems to be oscillating with either a very large period or a very small one.

```
1   1   4   4   8   8   8   1   1   4   4   1   1   1   1   1
1   1   1   1   8   8   4   1   1   1   1   1   1   1   4   1   1
4   1   1   1   8  72   4   1   1   1   1   1   1   1   4   1   4
4   1   1  72  72  36   4   1   2   1   1   1   1   1   1   1
4   4   1  18   1  18  36   2   2   2   1   1   1   1   6   1
4   1   1  18  18  18  36   1   1   1   1   1   1  12   6  12
12  1  18  18  18  18  18   1   1   1   1   1  12  12   1  12
1   1  18  18  18   1  18   1   1   1   1   1   1  12  12  12
1   1   1   1   1   1   1   1   1   4   4   1  12  12   1   1
1   1   1   1   1   1   1   1   1   1   4   4   4  12   1   1
1   4   4   1   1   1   1   1   1   1   4   4   1   4   1   4
1   1   1   1   1   1   1   1   1   1   1   4   1   4   4   4
1   1   1   1   2   2   1   1   1   1   1   1   1   1   4   1
1   1   1   1   1   1   1   1   1   1   1   1   1   1   4   1
1   1   1   8   8   1   1   1   1   1   1   4   4   1   1   1
1   1   8   8   8   8   1   1   1   1   4   1   1   4   4   1
```

Figure 5
Local periods in the
frozen regime
p = 0.22

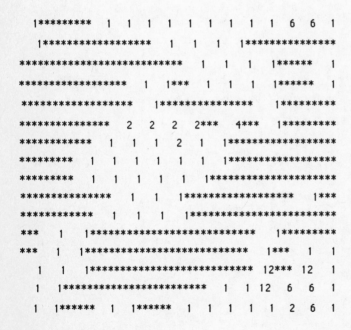

Figure 6
Local periods in the
chaotic regime
p = 0.30
(the stars correspond to
periods larger than 1000)

Percolation.

```
. 3 3 3 3 1 . 3 3 9 9 9 9 9 . . . . . . . 7 7 7 . . . 9 . . .
. . . 3 3 3 3 3 3 3 1 . 9 . . . . . . . . 7 7 7 7 . 9 9 7 . 2
2 . . . . 3 . 3 3 3 3 3 . . . . . . . . 9 9 . . 7 7 7 6 9 . .
. . . . 3 3 3 . . . 8 3 8 8 . 6 6 . . . 9 2 2 . 7 7 8 9 . .
. . . . 3 9 . . . 3 8 8 8 6 . . 1 . . . . 2 2 2 7 . 8 . . .
. . . . . 9 . . . 3 . . 8 . 5 5 3 . . . . . . . . 9 . . . .
. . . . . . . . . . . . . . 5 5 4 . . . . . . 4 4 . 9 9 9 9 .
. . . . . . . . . 2 . . 2 . . . 4 1 . 4 4 4 . 4 9 9 . . . .
. . . . . . . 9 9 . 2 2 2 2 . . . . 4 4 4 4 4 9 9 9 9 . . . 9
. 9 9 . . 9 9 9 . 2 2 . . . . . . . 4 4 . . 9 4 . . . . . 9
9 9 9 . . . 9 . 2 2 . . . . . . . . 4 . . . 9 9 . . . . . .
9 9 9 9 . . 9 . . 2 . . 4 4 . . . . 4 . . 9 9 . . . . . . .
9 9 9 . 9 9 7 . . . . 4 4 . . . . 4 9 . 9 9 . . . 9 . . 9
9 . . . . . . 2 . . . . . . . 9 9 9 9 8 8 3 . . . 9 . . 9
9 . . . 1 2 2 . . . 9 9 . . 9 9 9 9 9 8 8 . 1 1 . 9 9 . 9 9
. . 1 . 1 . . . . . 9 . . . 9 9 . 9 9 8 . . . . . 9 9 . . 9
9 . 1 1 1 . . 9 9 9 . . . 9 9 9 9 7 9 9 9 . . 9 9 9 . . 9
9 9 7 4 2 2 . . 9 . 9 9 . . . 9 9 9 9 9 9 9 . 9 9 . . . 1 1
. 9 9 9 . . . . 9 . 1 9 9 . . . . . 9 9 . 9 . 9 9 9 . . 5 .
. . 8 6 . . . . 9 . 1 9 . . . . . 9 9 . 9 9 . 9 9 9 9 4 5
5 3 5 2 2 . . . . . . . . . . . 9 . 9 9 9 . . . . . . 9 5
5 . . . . . . . . . . . . . . . . . . . . . . . . . . 9 5
5 . . . . . . . . . . 9 9 . . . . 9 9 9 . . . . . . . 9 9
. 9 9 . 9 9 . . . . . . 9 . . 9 9 9 9 . . . . . . . . .
. 9 9 9 9 . . . . . . . 9 9 . . . 9 9 . . . . . . . . 9
9 . 9 9 9 9 . . . . . . 9 . . . . . 9 9 . . . . . . . 9
9 . 9 9 9 . . . . . . 9 . 9 9 . . . 8 9 . . . . . . 9 9 9
9 9 9 9 . . 9 . . . 9 8 9 8 . . . . . . . . . . . . . 9 .
. . . 9 . . 9 . . . 8 8 8 . . . 9 9 9 9 9 9 9 . . . 9 9 .
. . . 3 . 9 9 9 3 . . . 9 9 9 . . . 9 . . 7 7 7 9 . . 9 9 9 .
```

Figure 7
Frozen regime
p=0.21

```
. . . . . . . . . 9 9 9 9 2 9 9 9 8 6 6 6 1 9 9 7 . . . .
. . 3 3 3 . . . 9 . . . . 9 9 9 9 9 9 6 5 6 6 6 . . . . . . 1
. . . . . . . . 9 9 9 . . 9 9 9 9 9 9 7 6 6 . . . . 9 . . . 1
. . . 9 . 3 9 9 9 . 9 9 9 9 9 9 9 9 8 8 6 2 . . . 9 9 9 9 .
. . . 9 3 3 3 . . . . 9 9 9 9 9 9 9 9 9 6 2 . . . 9 9 9 9 9
. . . 8 . . . . . . 9 9 9 9 9 8 9 8 9 8 9 9 9 9 9 9 8 9
. . . 1 1 1 1 1 9 9 9 9 9 9 8 9 8 9 8 5 8 9 9 5 9 9 9 9 9
1 . . . 1 1 1 1 1 8 8 9 9 9 9 9 9 9 9 9 9 9 6 9 9 9 . . 9 9
9 9 . . . . 3 3 3 2 9 9 9 3 3 6 9 7 9 9 9 5 2 9 9 9 . . 9 9
1 1 1 . . . 3 3 3 3 3 9 3 3 3 . . . . . 9 9 2 9 7 9 9 . . .
1 . 1 . . . 3 3 9 9 3 3 . 3 . . . . . . 9 9 . 9 9 . 9 9 .
. . . . . . 3 9 9 8 3 3 3 3 2 8 . . 9 9 9 9 9 9 9 9 9 9 9
9 . 9 9 . . . 1 1 9 9 3 3 3 2 8 . . . 9 9 . . 9 9 9 9 9 9 9
9 9 9 9 . . . 1 . 8 9 8 3 . 8 8 . . . 9 . . . . 9 9 9 . 9
9 9 9 9 . 9 9 9 9 9 8 8 8 8 1 1 . 5 . . . . . . . . . .
9 9 9 9 . 9 9 9 9 9 9 8 8 8 . 1 1 . 5 . . 9 9 9 . . . . .
9 7 9 9 9 . . 8 9 2 8 9 8 . . . . 8 . . 3 8 . . 2 7 2 2 .
9 9 . . 9 9 . . 9 8 8 8 9 8 8 . 9 9 9 9 3 9 . . 7 6 6 . 9
9 7 7 . . . . . 9 9 8 8 9 9 . 9 9 9 9 9 9 9 . . 6 6 9 9
9 9 . . . 9 9 9 9 8 9 8 8 9 . 9 9 9 9 9 9 8 . 9 6 6 . 4
. . . . 9 9 9 9 9 9 8 9 9 9 9 9 9 9 9 9 8 9 9 9 6 . 4
. . . . 9 9 8 9 9 7 8 9 9 9 9 7 . . 9 9 9 9 9 9 9 1 . 6 . .
. . 9 9 . 9 9 9 9 8 8 9 . . . . . . . 9 9 . . 9 . 1 3 . .
. . 9 . . 7 7 9 . 6 9 . . . . . . . . 9 9 . . 9 . 2 3 3 3 .
. . . . 7 7 7 2 9 9 1 1 . . . . . . . . . . . 9 . 2 3 3 .
. . . . 7 7 7 3 9 . 2 . . . . . . . . . . . . 1 . 3 9 3 3 .
. . . . . . . . 2 2 2 . . 9 9 . . . . 9 . 1 1 1 3 9 . . 9
9 9 . . 3 . . . . . . . 9 2 2 2 9 9 9 9 . 4 4 . 1 3 . . .
. 9 9 . 3 . . . 9 9 9 9 9 9 . 9 9 9 9 1 1 1 4 1 9 9 . 9 9 .
. 9 9 . . . . 9 9 9 9 9 9 9 9 9 9 9 9 6 6 6 3 2 7 7 7 . . .
```

Figure 8
Chaotic regime
p = 0.28

Figures 9 and 10 are histograms of oscillations equivalent to figure 3. They show how many times out of 9 initial conditions each automaton is oscillating during the limit cycle. For small values of p, the oscillating regions are small clusters separated by the stable core. For larger p, the oscillating cluster percolate through the sample. Computer simulations show that the percolation threshold is p = 0. 26 ± .02 which is the same value for the frozen/chaotic transition as determined by other methods (see further).

Evolution of the overlaps.

As in the case of continuous systems we expect some sort of strong sensibility to the initial conditions in the chaotic behavior. In order to compare trajectories in the phase space, one computes the overlap between successive configurations defined as the ratio of the number of automata which are in the same state to N. If starting from some initial condition a few automata are flipped, the evolution of the overlap between the two configurations (the perturbed and unperturbed configurations) indicates whether they converge to the same attractor, whether they remain at some distance proportional to their initial distance or whether they diverge as in continuous chaotic dynamics. Figure 9, from Derrida and Stauffer [5], compares d_∞ the distances at "infinite time" as a function of the initial distance for four input cellular nets. In the chaotic regime the relative distance evolves towards a finite value of the order of 0.1, however small the initial perturbation is. In the frozen regime d_∞ is proportionnal to d_0: because the frozen behavior corresponds to independently oscillating subnets, for small value of d_0, only a few subnets are perturbed in proportion to d_0 and d_∞ varies accordingly. Derrida and Stauffer obtained p_c=0.26 for the transition threshold by plotting d_∞ as a function of p for fixed d_0.

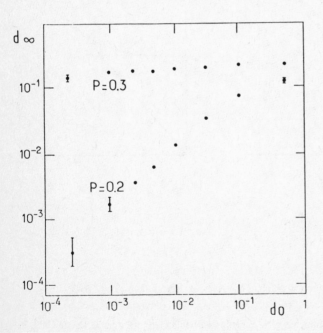

Figure 9 Distances at 'infinite time' versus initial distances, in the frozen (p=0.2) and the chaotic regime (p=0.3).

Annealed nets.

In the case of Kauffman nets of random connectivity k, the evolution in one time step of the overlap x(t) between two random configurations can be predicted. x^k is the proportion of automata whose k inputs are in the same state for both configurations. These automata will be in the same state at the following time step and all the other automata will be in the same state with probability 1/2. x then varies as:

$$x(t+1) = \frac{1 + x^k(t)}{2}$$

Such an expression is only valid for random configurations and, in principle, cannot be indefinitely iterated, which would give us interesting indications about the infinite time behavior of the system. One way to get rid of this difficulty is to invent a new type of automaton which function is randomly changed at each time step: these "annealed" nets have been proposed by Derrida and Pomeau [12]. In this case the above expression can be iterated. Two types of iterated maps exist: for k less than or equal to 2, x goes to 1 (identical configurations) at infinite time for infinitely large nets. This implies that the volume of the configurations space available to the system goes to 0 and that an organized behavior has been reached. This is not the case for k larger than 2 (see figure 10).

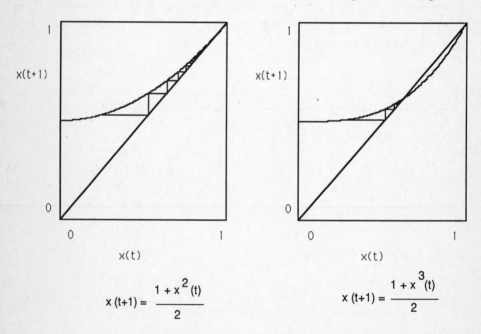

$$x(t+1) = \frac{1 + x^2(t)}{2} \qquad\qquad x(t+1) = \frac{1 + x^3(t)}{2}$$

Figure 10 Iteration graph of the relative overlap for k=2 and k=3.

We have done computer simulations both for annealed nets and for normal Kauffman nets (Derrida and Weisbuch [9]). Surprisingly enough annealed and deterministic ("quenched") nets exhibit the same behavior (see figure 11), except at large time for k=2. In this latter case scaling

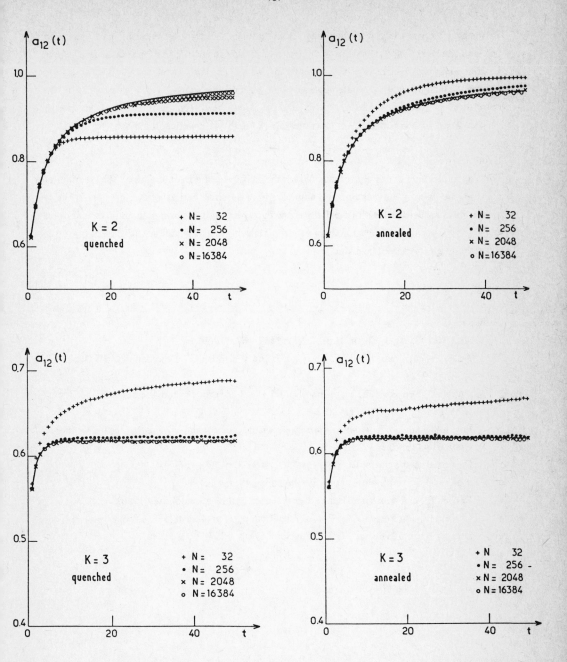

Figure 11. Comparison of the time evolutions of the overlaps between two configurations for quenched and annealed nets, for connectivity k=2 and 3. The continuous line is obtained by iterating the relative overlap equation equation.

effects are observed which show that the overlap saturates at intermediate times. This similarity in behavior can be explained. Modeling a deterministic net by an annealed net on several time steps is valid as long as the ancestors of an automaton are all different - by ancestor we mean all the automata which influence the state of an automaton after t time steps-. In such a case there are no correlations between the inputs of a given automaton and the analytic expression can be iterated. This approximation is thus valid for time intervals which increase with the size of the net.

Conclusions.

Pattern formation is definitely responsible for the existence of a frozen regime. The isolation of small patches of oscillating automata is at the origin of the small periods of the attractors. It is also responsible for the robustness of the dynamics with respect to small changes in initial conditions or in the transition rules of a few automata. It is the basis of the interpretation of the attractors of the dynamics as an "organized behavior".

Bibliography.

1   Kauffman S. A., J. Theor. Biol., **22**, pp. 437-467, (1969).

2   Atlan H., Fogelman-Soulié F., Salomon J. and Weisbuch G., Cybernetics and Systems,**12**, p.103, (1982).

3   Atlan H., Ben-Ezra E., Fogelman-Soulié F., Pellegrin D. and Weisbuch G., J. theor. Biol.,**120**, pp. 371-380, (1986).

4   Fogelman-Soulié F., *Contribution à une théorie du calcul sur réseau*, Thesis, Grenoble University (1985).

5   Derrida B. and Stauffer D., Europhysics Letters **2**, p. 739, (1986).

6   Weisbuch G. and Stauffer D., J. de Physique, **48**, p.11,(1987).

7   Derrida B., in *Chance and Matter,* Les Houches Summer School, July 1986.

8   Derrida B., and Pomeau Y., Europhysics Letters, **1**, pp.45-49, (1986).

9   Derrida B., and Weisbuch G., J. Physique,**47**,pp. 1297-1303, (1986).

COMPLEXITY IN ECOLOGICAL SYSTEMS

Jean-Arcady MEYER

Groupe de BioInformatique. CNRS UA686.

Departement de Biologie. ENS. Paris.

If opinions do indeed vary as to what exactly is covered by the notion of complexity, it would nevertheless appear that complexity is invariably construed as an obstacle to comprehension (ATLAN, 1986).

In this sense, it is in the first place indissolubly linked in the eyes of the ecologist with the existence of a multiplicity of variables which have to be accounted for if he is to succeed in understanding the operating laws of any given ecological system.

When these variables correspond to the static description of a situation - for example in the form of a table of measurements of m variables carried out on n elements - a series of effective methods are available to cope with the complexity of the information involved, methods which are included under the generic term of "multivariate data analysis" (COOLEY and LOHNES, 1971). Among these, the "principal components analysis", for example, makes it possible to replace the original m variables, which are more or less intercorrelated, by a lesser number of independant variables, while preserving the largest possible portion of the original information.

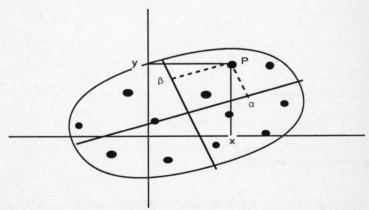

Figure 1. An element like P is characterized by its coordinates x and y in the space of the original variables, as well as by its coordinates α and β in the space of the principal components.

To be more specific, this technique entails a substitution, for the m original variables, of the p primary axes of the hyperellipsoid formed by the m elements represented in the space of these m variables. These elements can be characterized by their coordinates within the original m variable space, as in that of the p principal components which were retained (Figure 1).

Thus these latter have the same status as the original variables: they represent hypothetical factors determining the respective positions of the elements, factors that will need to be interpreted in explaining these positions and their significations.

An example of the application of such technique is given in the work of LAVELLE(1986), who has demonstrated that a large number of variables defining the environment and the biology of several communities of earthworms could be summarized by a single factor, which represented essentially the mean temperature of the environmental context in question. These communities can be divided in five ecological categories: epigeic, anecic, oligo-, meso- or polyhumic endogeic.

The proportions of these categories in each community, when plotted against environmental temperature (Figure 2), reveal the narrowness of the ecological niche occupied by certain species suggesting that, with increasing temperatures, earthworms become able to exploit increasingly poorer organic resources, since their mutualistic associations with soil microorganisms become more and more efficient.

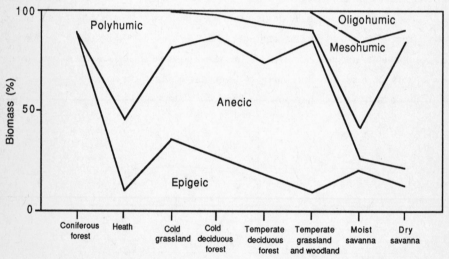

Figure 2. Trophic structures of earthworm communities in relation to the mean environmental temperature.

In this perspective, it is clear that the number of independant factors that must be accounted for when explaining a given reality is a measure of the complexity of the system involved.

In recent years, a dynamic equivalent to the preceding approach has been developed. Although the potential dynamic complexity of ecological systems has indeed been widely recognized beginning in the early 1970's and although considerable interest has focused on non-linear systems since that time (MEYER, 1978), it is nonetheless true that few natural ecological systems had been studied to date with the purpose of identifying possible strange attractors.

The main reason for this is clearly in relation to the problem previously mentioned, that of the large number of state variables characterizing the dynamics of any ecosystem - such as the numbers of individuals belonging to the species involved. Generally these numbers cannot be quantified and frequently the species concerned are not even known. It accordingly seemed impossible to trace the dynamics of the system in its phase space and thus to gain any knowledge whatsoever of the complexity of its dynamics.

Now, however, in the light of a proposal by TAKENS(1981), it has become possible to reconstitute a given phase space by monitoring a single species belonging to the system under study. In order to do this, all one needs to know is the number of individuals at regular time intervals: t=s, s+T, s+2T... s+(m-1)T. It appears that, for virtually every X variable, the m-dimensional graph obtained from points with coordinates X(s), X(s+T)... X(s+(m-1)T) displays the same dynamic properties as the phase space derived from the n variables associated with the species of the original system. A sufficient condition is that m > 2n+1. It is, however, not necessary and experience shows that an m value much lower than n may be sufficient.

In SCHAFFER(1984), an application of this "minimum embedding" technique to the study of the well-known lynx/hare cycle in Canada is to be found. A large amount of data on this cycle is available, in relation with fur trade statistics for the period 1735 to 1940. It seems that this cycle is determined by the existence of a three-dimensional strange attractor and therefore that at least three species - lynx, rabbit, and vegetation for instance - play a part in the phenomena observed. These results thus support the findings of GILPIN(1973), who concluded that the known data were incompatible with a two-species predator-prey model.

Whatever the case, it is obvious that the embedding process just mentioned does indeed reflect the complexity of a given ecological system , as do also the various measures commonly associated with strange attractors, notably their fractal dimension and Lyapunov exponents.

Another obstacle to understanding the operating laws of ecological systems is often related to the fact that the ecologist does not always choose the most efficient world-view. For instance, he may persist in seeking to interpret in mechanistic terms a given aspect of the way the system functions, whereas adaptive terms would be more suitable, embedding the system in question in an evolutionary perspective (MEYER, 1982). It should moreover be noted that these two approaches do not have the same objective, the former aiming to answer questions as to the "how" of the phenomena, while the latter is concerned whith "why"-type questions.

A study by NOLLET(1988) devoted to the determinism of the queen's egg-laying in a colony of bees gives a clear illustration of these points. All attempts at establishing a functional relationship between the egg-laying rate and a variety of physical parameters of the environment - such as sunlight or the availability of nectar - in view of predicting the variations in this factor had indeed proven vain. In particular, no mechanistic dependency of this type enabled one to account for the fact that, under certain conditions, an increase in egg-laying actually precedes an increase in available environmental resources. On the other hand, if one goes on the hypothesis that the bee colonies observed today are the product of a very long-term selection resulting in an optimization of their chances of survival, it becomes possible to make explicit use of such an optimization hypothesis within a dynamic model and to come up with the anticipatory effects just alluded to.

The relationship is evident between such a result and the experiments described in LOUVEAUX(1965). It was, in effect, observed that when hives are transplanted from one environmental context to another the queens persist in their previous egg-laying pattern in the new context because they are still subject to the program perpetuated in their genome through natural selection.

A connection between the notion of optimization and that of complexity is brought up in a study of ULANOWICZ(1986). Using the various compartments that can be delimited within an ecological system and the energy flows passing through them, this author defines a variable he calls "internal ascendency" that measures how well, on the average, the system articulates a flow event from any one compartment to effect any other specific compartment. In ULANOWICZ's opinion, all natural systems are inclined to grow and develop in such a way as to maximize this variable, a concept to be compared with other optimization hypotheses relative to ecosystems, for example those of CONRAD(1972), of JORGENSEN and MEIER(1979) and of ODUM(1983).

If we use T_{ji} to represent the flow between compartments j and i, T_j to represent the sum of flows issuing from compartment j, and T'_i to represent the sum of flows entering compartment i, the internal ascendency of the system, supposed to be in steady state, can be defined as:

$$A_I = T \sum^n \sum^n f_{ji} Q_j \log(f_{ji}/Q'_i)$$

where

$$n = \text{number of compartments,}$$
$$f_{ji} = T_{ji}/T_j$$
$$Q_j = T_j/T$$
$$Q'_i = T'_i/T$$

The quantity T indicates the total throughput, that is the total of all the flows entering, passing through, and exiting from the system.

It can be demonstrated that the above expression can be rewritten in the following form :

143

$$A_I = -T \sum Q_j \log Q_j - [-T \sum e_j Q_j \log Q_j \ -T \sum r_j Q_j \log Q_j \ -T \sum \sum f_{ji} Q_j \log(f_{ji} Q_j / Q'_i)]$$

with terms e_j and r_j standing respectively for the amount of effective and dissipated energy leaving the system from a given compartment j.

Under these conditions, the internal ascendency is seen to be less than or equal to the quantity

$-T \sum q_j \log Q_j$, called "internal development capacity". This quantity is limited by the three bracketed terms which indicate what proportion of the energy entering the system is exported, dissipated, or used to insure the "functional redundancy" of the system, that is the ambiguity of its internal connections .

For example, in the case of the Cone Spring ecosystem described by TILLY(1968) (Figure 3), the following values were computed for the parameters described above:

Total throughput	= 42445 Kcal $m^{-2} y^{-1}$
Internal development capacity	= 71371.577 Kcal bits $m^{-2} y^{-1}$
Internal ascendency	= 29331.977 Kcal bits $m^{-2} y^{-1}$
Exported energy	= 2971.333 Kcal bits $m^{-2} y^{-1}$
Dissipated energy	= 28557.946 Kcal bits $m^{-2} y^{-1}$
Functional redundancy	= 10510.320 Kcal bits $m^{-2} y^{-1}$

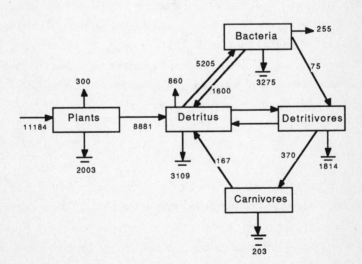

Figure 3. Energy flows (Kcal $m^{-2} y^{-1}$) in the Cone Spring ecosystem. Arrows not originating from a compartment represent inputs from outside the system. Arrows not terminating in a compartment represent exports of useable energy out of the system. Ground symbols represent dissipations of respired energy.

One may consider that the energy exported outside the system is used in the organization and maintenance of the superstructures of which this system is a constituent, whereas the dissipated energy, in like manner, is used for the organization and maintenance of substructures within each compartment characteristic of this system.

One may further consider that the three bracketed terms in the preceding expression reflect the degree of complexity of the system in question. They indeed represent a conditional entropy measuring the uncertainty remaining after the flow structure has been specified. This entropy is a function both of T and n - thus of the system size - and of the quantities e_j, r_j, f_{ji}, Q_j and Q'_i - thus of its structure.

Therefore, according to ULANOWICZ, the evolution of any ecological system would be the result of the search for the best compromise between two opposite tendancies, the first being to grow, the second to increase both its functional redundancy and the firmness of its attachment in the inferior and superior levels of integration.

Here is another reason why ecological systems are complex and difficult to elucidate: the necessity of introducing into the corresponding reasonings every level of integration characteristic of ecology, namely the individual, the population, the community, the ecosystem and the landscape.

In conclusion, it is clear that ecological systems are complex because they are characterized by a large number of intercorrelated variables and because they generally display highly non-linear dynamics. The laws that govern their present-day operation may be strongly dependent upon their past, as they may likewise be dependent upon the laws of the sub-systems of which they are made up or of the supersystems of which they are a part.

Although certain measures of the complexity of ecological systems have been presented in this text, they only address one or the other of these aspects. Great advances in ecological theory can be expected to derive from measures that would integrate them all.

REFERENCES

ATLAN.H. 1986. La complexite naturelle et l'auto-création du sens. In AIDA.S. and al. Science et pratique de la complexite : actes du Colloque de Montpellier. Mai 1984. La Documentation Francaise.
CONRAD.M. 1972. Statistical and hierarchical aspects of biological organization. In WADDINGTON.C.H. Towards a theoretical biology, Vol 4. University Edinburgh Press. pp189-220.
COOLEY,W.W and LOHNES,P.R. 1971. Multivariate data analysis. Wiley.
GILPIN,M.E. 1973. Do hares eat lynx? Amer. Natur. 107, 727-730.
JORGENSEN,S.E. and MEIER,H. 1979. A holistic approach to ecological modelling. Ecol. Model. 7, 169-189.
LAVELLE,P. 1986. Associations mutualistes avec la microflore du sol et richesse specifique sous les tropiques: l'hypothese du premier maillon. C.R.Acad.Sci. Paris.302,Serie III, 1, 11-14.

LOUVEAUX,J. 1966. Les modalites de l'adaptation des abeilles (Apis mellifica L.) au milieu naturel. Ann. de l'Abeille. 9, 323-350.

MEYER,J.A. 1978. Sur la dynamique des systemes ecologiques non lineaires. J. Phys. 39,8, 29-37.

MEYER,J.A. 1982. Les modeles de simulation de la dynamique du plancton: nature, utilisation et limites. In POURRIOT,R. Ecologie du plancton des eaux continentales. Masson. pp 147-193.

NOLLET,P. 1988. Contribution a l'etude des strategies adaptatives des insectes sociaux. These de l'Universite Paris 7. Specialite Biomathematiques.

ODUM,H.T. 1983. Systems ecology. An introduction. Wiley.

RUTLEDGE,R.W., BASORE,B.L. and MULHOLLAND,R.J. 1976. Ecological stability: an information theory viewpoint. J.Theor.Biol. 57, 355-371.

SCHAFFER,W.M. 1984. Stretching and folding in lynx fur returns: evidence for a strange attractor in nature? Amer. Natur. 124, 798-820.

TAKENS,F. 1981. Detecting strange attractors in turbulence. In RAND,D.A. and YOUNG,L.S. Dynamical systems and turbulence. Springer-Verlag. pp 366-381.

TILLY,L.J. 1968. The structure and dynamics of Cone Spring. Ecol. Monographs. 38, 169-197.

ULANOWICZ,R.E. 1986. Growth and development. Ecosystems phenomenology. Springer-Verlag.

LIST OF PARTICIPANTS

M. Ageno
Dipartimento di Fisica,
Universita' La Sapienza,
Piazzale Aldo Moro, 2
I-00185 ROMA (Italy)

P.Ambrosino
Dipartimento di Fisica
Gruppo Biofisica
Mostra d'Oltremare - Pad.16
Via Giovenale 1
80122 NAPOLI

D.J. Amit
Racah Institute of Physics,
Hebrew University,
91904 Jerusalem, (Israel)

C.Arcelli
Istituto di Cibernetica - C.N.R.
Via Toiano, 6
80072 ARCO FELICE (NA)

H. Atlan
Department of Medicine,
Biophysical and Nuclear Medicine
Hadassah Medical University,
Kiryat Hadassah
Jerusalem (Israel)

C. Bachas
Centre de Physique Theorique,
Ecole Polytechnique
91128 Palaiseau Cedex
(France)

G.Battimelli
Dipartimento di Fisica
Università di Roma "La Sapienza"
P.le A.Moro, 2
00185 ROMA

R.Biancastelli
Dipartimento di Fisica
Università di Roma "La Sapienza"
P.le A.Moro, 2
00185 ROMA

C.Boldrighini
Dipartimento di Matematica
Università degli Studi
CAMERINO (MC)

D.P. Bovet
Dipartimento di Matematica,
Universita' "La Sapienza"
Piazzale Aldo Moro 2,
I-00185 ROMA

E.Caglioti
Fondazione U.Bordoni
Via G.G.Porro, 22
ROMA

P.Campadelli
Istituto di Fisiologia dei Centri Nervosi
C.N.R.
Via M.Bianco, 9
MILANO

R.Campanella
Dipartimento di Fisica
Università di Roma "La Sapienza"
P.le A.Moro, 2
00185 ROMA

P.Carnevali
I.B.M.
Via Giorgione, 159
00147 ROMA

M.Cassandro
Dipartimento di Fisica
Università di Roma "La Sapienza"
P.le A.Moro, 2
00185 ROMA

E.Castelli
Ecole des Hautes Etudes Sciences Sociales

F.Catoni
ENEA - CRE Casaccia
Via Anguillarese
ROMA

C.Cilli
Istituto di Filosofia
Via Prato della Signora, 15
00199 ROMA

M.Cini
Dipartimento di Fisica
Università di Roma "La Sapienza"
P.le A.Moro, 2
00185 ROMA

B. Continenza
Dipartimento di Ricerche Filosofiche
Università di Tor Vergata
Via O.Raimondo
00173 ROMA

L.Contoli
Centro Genetica Evoluzionistica
C.N.R.
Via Lancisi, 29
00161 ROMA

R.Conversano
ENEA Casaccia
Dir. Centrale Studi

G.Corbellini
Scuola Superiore di Storia della Scienza
Domus Galilaeana
Via Selvareggia, 144
CADEO (PC)

G.Cosenza
Dipartimento di Fisica
Università di Napoli
Mostra d'Oltremare - Pad.19
80125 NAPOLI

A.Crisanti
RACAH Institute of Physics
Hebrew University
JERUSALEM (Israel)

N.Cufaro Petroni
Dipartimento di Fisica
Via Amendola, 173
70100 BARI

C.D'Antoni
Dipartimento di Matematica
Università di Roma "La Sapienza"
P.le A.Moro, 2
00185 ROMA

G.F.De Angelis
Dipartimento di Fisica
Università
SALERNO

A.De Luca
Dipartimento di Matematica
Università di Roma "La Sapienza"
P.le A.Moro, 2
00185 ROMA

F.De Pasquale
Dipartimento di Fisica
Università di Roma "La Sapienza"
P.le A.Moro, 5
00185 ROMA

D.Dohrn
Dipartimento di Fisica
Università
SALERNO

S.Forestiero
Dipartimento di Biologia
Università di Roma, Tor Vergata
Via O.Raimondo
00173 ROMA

E.Gabrielli
Dipartimento di Fisica
Università di Roma "La Sapienza"
P.le A.Moro, 2
00185 ROMA

P.G.Gabrielli
ENEA - CRE Casaccia
Via Anguillarese, 301
ROMA

A.Giansanti
Dipartimento di Fisica
Università di Roma "La Sapienza"
P.le A.Moro, 2
00185 ROMA

P. Grassberger
Department of Physics,
University of Wuppertal,
Gauss-str. 20,
D-56 WUPPERTAL
(Germ. Federale)

Z. Grossman
Faculty of Medicine
Tel-Aviv University
TEL AVIV (Israel)

F.Guerra
Dipartimento di Fisica
Università di Roma "La Sapienza"
P.le A.Moro, 2
00185 ROMA

H.Hohewegger
Via A.Scarenzio, 13
00123 ROMA

G.Immirzi
Dipartimento di Fisica
Università di Napoli
Mostra d'Oltremare - Pad.19
NAPOLI

F.Koukiou
Institute de Physique Theorique
1015 Lausanne (Suisse)

U.Krey
Institut fur Physik III des Universitat
Universitat Regensburg, Physics Dept.
D-84 Regensburg

R.Livi
Dipartimento di Fisica
Università di Firenze
L.go E.Fermi, 2
50125 FIRENZE

M.Marinucci
ENEA - CRE Frascati
Div.Fusione
00044 FRASCATI

G.F.Mattioli
Dipartimento di Matematica
Università di Roma "La Sapienza"
P.le A.Moro, 2
00185 ROMA

P.Mentrasti
Dipartimento di Matematica
Università di Roma "La Sapienza"
P.le A.Moro, 2
00185 ROMA

J.A. Meyer
Departement de Zoologie,
Ecole Normale Superieure,
rue d'Ulm,
F-75231 PARIS Cedex 05 (Francia)

A.Morelli
SISSA
Strada Costiera, 11
TRIESTE

J.P. Nadal
Groupe de Physique Theorique,
Ecole Normale Superieure,
24, rue Lhomond
F-75231 PARIS Cedex 05 (Francia)

C.Nardone
ENEA - CRE Frascati
00044 FRASCATI

S.Nicolis
Laboratoire Physique des Solides
Universitè Paris-Sud
91405 ORSAY (Cedex)

P. Omodeo
Dipartimento di Biologia
Universita' di Roma "Tor Vergata"
Via O.Raimondo
00173 ROMA (Italy)

G.V.Pallottino
Dipartimento di Fisica
Università di Roma "La Sapienza"
P.le A.Moro, 2
00185 ROMA

G. Parisi
Dipartimento di Fisica,
Universita' di Roma "Tor Vergata"
Via O.Raimondo
00173 ROMA (Italy)

L.Peliti
Dipartimento di Fisica
Università di Roma "La Sapienza"
P.le A.Moro, 2
00185 ROMA

P.Pede
Via Bacchiglione, 3
00199 ROMA

D.Pescetti
Dipartimento di Fisica
Università di Genova
Via Dodecaneso, 33
16146 GENOVA

D.Petritis
Institut de Physique Theorique BSP
Universitè de Lausanne,
CH-1015 Lausanne (Switzerland)

G.Poppel
Inst. fur Physik III, Lst. Hoffmann
Universitat Regensburg
D-84 Regensburg

R.Rechtman
Dipartimento di Fisica
Università di Firenze
L.go E.Fermi, 2
50125 FIRENZE

A. Rossi
Dipartimento di Fisica
Università di Roma "La Sapienza"
P.le A.Moro, 2
00195 ROMA

G.C.Rossi
Sezione di Roma INFN e
Dipartimento di Fisica
Università
L'AQUILA

S.Ruffo
Dipartimento di Fisica
Università di Firenze
L.go E.Fermi, 2
50125 FIRENZE

M.Scalia
Università di Roma "La Sapienza"
P.le A.Moro, 2
00195 ROMA

E.Scoppola
Dipartimento di Fisica
Università di Roma "La Sapienza"
P.le A.Moro, 2
00195 ROMA

G.B.Scuricini
ENEA - CRE Casaccia
SP. Anguillarese 301
00100 ROMA

V. Somenzi
Cattedra di Filosofia della Scienza
Istituto di Filosofia
Villa Mirafiori
Via Carlo Fea 2
00161 ROMA

U.Stiegler
Max-Planck - Inst. fur Physik
Fohringer Ring 6
D-8000 Munchen

M.Testa
INFN Sezione di Roma e
Dipartimento di Fisica
Università
LECCE

G.Tratteur
Dipartimento di Fisica
Mostra d'Oltremare - Pad.19
NAPOLI

L.Vincenzotti
ENEA - DISP ARAMED

A.Vulpiani
Dipartimento di Fisica
Università di Roma "La Sapienza"
P.le A.Moro, 2
00185 ROMA

G. Weisbuch
Groupe de Physique Theorique
Ecole Normale Superieure,
24, rue Lhomond
F-75231 PARIS Cedex 05 (Francia)

Y.Cheng Zhang
Dipartimento di Fisica
Università di Roma "La Sapienza"
P.le A.Moro, 2
00185 ROMA

J.Zittartz
Institut fur Theoretische Physik
Universitat Koln
Zulpicher str. 36
5000 KOLN 41 (W.Germany)

Lecture Notes in Mathematics

Lecture Notes in Physics